ALPHA CENTAURI

SIRIUS

HT YEARS OF THE SUN

SCUTUM-CENTAURUS ARM

GALAXY DISK

ORION S

EINSTEIN'S SHADOW

Also by Seth Fletcher

Bottled Lightning:
Superbatteries, Electric Cars, and the New Lithium Economy

EINSTEIN'S SHADOW

A BLACK HOLE, A BAND OF ASTRONOMERS, AND THE QUEST TO SEE THE UNSEEABLE

SETH FLETCHER

AN IMPRINT OF HARPERCOLLINS*PUBLISHERS*

Elements of this book were previously published in a different format in *Popular Science*.

HarperCollins books may be purchased for educational, business, or sales promotional use. For information, please email the Special Markets Department at SPsales@harpercollins.com.

FIRST EDITION

Designed by Suet Yee Chong
Title page image by 4Max/Shutterstock, Inc.
Illustrations and endpaper image courtesy of Katie Peek

Library of Congress Cataloging-in-Publication Data has been applied for.

ISBN 978-0-06-231202-0

18 19 20 21 22 LSC 10 9 8 7 6 5 4 3 2 1

For Sylvia

Heaven knows what seeming nonsense may not tomorrow be demonstrated truth.

—ALFRED NORTH WHITEHEAD

AUTHOR'S NOTE

We live twenty-six thousand light-years from the center of the Milky Way. That's a rounding error by cosmological standards, but still—it's far. When the light now reaching Earth from the galactic center first took flight, people were crossing the Beringian land bridge, hunting woolly mammoths along the way.

The distance hasn't stopped us from drawing a fairly accurate map of the heart of the galaxy. We know that if you travel inbound from Earth at the speed of light for about twenty thousand years, you'll encounter the galactic bulge, a peanut-shaped structure thick with stars, some nearly as old as the universe. Several thousand light-years farther in, there's Sagittarius B2, a cloud a thousand times the expanse of our solar system containing silicon, ammonia, doses of hydrogen cyanide, dashes of ethyl formate, which tastes like raspberries, and at least ten billion, billion, billion liters of alcohol. Continue inward for another three hundred and ninety light-years or so and you reach the inner parsec, the bizarro zone within about three light-years of the galactic center. Tubes of frozen lightning called cosmic filaments streak the sky. Bubbles of gas memorialize ancient star explosions. Glowing streams of gas arc toward the core. Gravity becomes a foaming sea of riptides. Blue stars that make our sun look like a marble slingshot by at millions of miles per hour. Space becomes a bath of radiation; atoms dissolve into a fog of subatomic particles. And near the core, that fog forms a great glowing Frisbee encircling a vast dark sphere. This is the supermassive black hole at the center of the Milky Way, the still point of our slowly rotating galaxy. We call it Sagittarius A*.

Every object in the Milky Way orbits the galactic center and, thus, the black hole, which is about as wide as Mercury's orbit around the sun. Our home star completes a circuit about every two hundred million years. Every galaxy probably has an enormous black hole at its core. Galaxies and their central black holes seem to evolve together. They go through stages. Sometimes the black hole spends eons inhaling matter as fast as physically possible, converting that matter into energy in a long-lasting cataclysm, each instant the equivalent of billions of thermonuclear weapons detonating simultaneously. In these "active" stages, the black holes fire jets of matter and energy across the universe, landscaping the cosmos like great rivers cleave continents and build deltas. Depending on their mood, black holes decide when their host galaxies can grow new stars: when they're on a rampage, blowing shock waves and howling cosmic winds, baby stars can't grow. When a black hole settles down into a quiescent state, the next generation of stars gets to form.

No one is sure how the black holes themselves formed. Astronomers have discovered black holes at the edge of the visible universe that contain the mass of billions of suns. These black holes must have reached this size when the universe was less than a billion years old. Yet according to the conventional understanding of how black holes grow, they *couldn't* have gotten so big so quickly. There hadn't been enough time. And yet, there they are.

In the fifty years since the physicist John Wheeler popularized the term "black hole," these objects have given people a lot to think about. They're strange enough to inspire buttoned-down scientists to earnestly ponder outlandish questions. Are we living inside a black hole? Was the Big Bang the flip side of a black hole forming in another universe? Does each black hole contain a baby universe? Can we use black holes for time travel? Would the near vicinity of black holes be a good place to look for extraterrestrial life?

Will scientists ever discover the most fundamental laws of

nature—a theory of everything? Black holes might be the key. The twentieth century produced two spectacularly successful theories of nature: general theory of relativity, and quantum theory. General relativity says the world is continuous, smoothly evolving, and fundamentally *local*: influences such as gravity can't travel instantaneously. Quantum theory says the world is twitchy, probabilistic, and nonlocal—particles pop in and out of existence randomly and seem to subtly influence one another instantly across great distances. If you're a scientist who wants to dig down to the deepest level of reality, the obvious question is: which is it?

General relativity describes the universe on the largest scales. Quantum mechanics governs the subatomic world. The two theories crash together most violently in black holes. We say, for example, that Sagittarius A* is a four-million-solar-mass black hole, implying that the black hole "contains" four million suns' worth of matter. But Einstein's equations tell us that the interior of a black hole is a vacuum, and that all the matter that has ever fallen in is packed into an infinitely dense surface at the center of the black hole called a singularity. To understand what happens at the singularity—and, by extension, at that other singularity known as the Big Bang—scientists need a theory of quantum gravity: a framework that unites general relativity with quantum mechanics.

It would be a lot easier to knit the two theories together if scientists could find loose threads to pull. Problem is, both quantum mechanics and general relativity have passed every experimental test they've ever been subjected to. But general relativity has never been tested near a black hole, where gravity conjures its true strength. And that's one of the many reasons scientists have long wanted to get a close look at a black hole.

It's funny, actually, how confidently scientists talk about black holes, given that no one has ever seen one. They've spent decades building mathematical models and observing the indirect effects of invisible masses that they've decided can only be black holes—but no one has ever gotten a direct look. If you could study a black hole up close, you could test the predictions that have piled up over the decades. Take Sagittarius A*, which is huge and, by cosmic standards, nearby, and therefore the best candidate for up-close study. A close look at Sagittarius A* could answer a long list of hard questions. Here's one example: general relativity predicts that Sagittarius A* will cast a shadow with a very specific shape. If astronomers get a picture of that shadow and it doesn't look the way they expect, they've just uncovered a major clue in the search for a deeper understanding of nature. It would be strong evidence that Einstein's equations are only an approximation of some deeper physical law—and it would provide clues about the identity of that deeper law. And if scientists ever come to understand nature at its most fundamental, it would be, as the late Stephen Hawking once wrote, "the ultimate triumph of human reason—for then we should know the mind of God."

This is a book about a group of astronomers on a quest to take the first picture of a black hole. They call their effort the Event Horizon Telescope. Their target is Sagittarius A*.

The book is based on nearly six years of reporting, starting in February 2012, with deep access to the project. I followed the astronomers to telescopes for tests and observing runs, attended their conferences, sat in on meetings, hung around their offices, stayed at their houses, and conducted so many interviews in person and via phone, email, Skype, Zoom, and text message that I'm not sure how to count them. With rare exceptions, I was the only journalist on the scene.

When I set out to follow the scientists of the Event Horizon Telescope (EHT) as they assembled the worldwide network of radio telescopes it would take to see to the edge of Sagittarius A*, I was confident I'd find a good story. The science was fascinating, the people were interesting, and the scenery—mountaintop observatories in places like Hawaii and Mexico—was excellent. But it didn't really click together until one night when I sat up talking to a wise astronomer at a hotel bar. We were at a weeklong conference, and a few of the EHT scientists had spent the evening pacing around a high table in the bar, sighing and muttering about the organizational chart they were trying to finish. They'd all gone back to their rooms by the time this astronomer explained to me what was really happening. "You know what they're fighting about, don't you?" he said. "They're fighting over who gets their name on the Nobel Prize."

THE OBSERVATORIES OF THE EVENT HORIZON TELESCOPE
A
B
C
D
E
F
G
H
BASELINES USED IN THE APRIL 2017 OBSERVATIONS
OTHER EVENT HORIZON TELESCOPE BASELINES

SMA
SUBMILLIMETER ARRAY
8-ANTENNA INTERFEROMETER
MAUNA KEA, HAWAI'I
ELEVATION 13,300 FEET

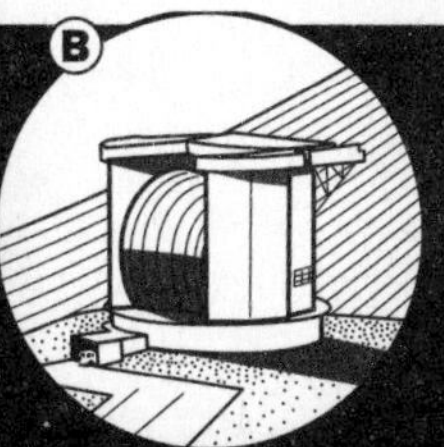

JCMT
JAMES CLERK MAXWELL TELESCOPE
SINGLE DISH
MAUNA KEA, HAWAI'I
ELEVATION 13,400 FEET

CARMA
COMBINED ARRAY FOR RESEARCH IN MILLIMETER-WAVE ASTRONOMY
23-ANTENNA INTERFEROMETER
CEDAR FLAT, CALIFORNIA
ELEVATION 7,300 FEET

SMT
SUBMILLIMETER TELESCOPE
SINGLE DISH
MT. GRAHAM, ARIZONA
ELEVATION 10,500 FEET

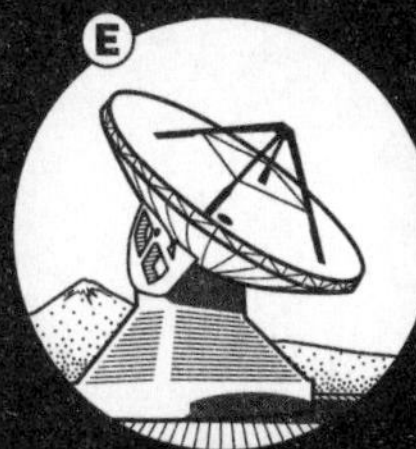

LMT
LARGE MILLIMETER TELESCOPE
SINGLE DISH
SIERRA NEGRA, MEXICO
ELEVATION 15,100 FEET

APEX
ATACAMA PATHFINDER EXPERIMENT
SINGLE DISH
ATACAMA DESERT, CHILE
ELEVATION 16,700 FEET

ALMA
ATACAMA LARGE MILLIMETER ARRAY
66-ANTENNA INTERFEROMETER
ATACAMA DESERT, CHILE
ELEVATION 16,400 FEET

SPT
SOUTH POLE TELESCOPE
SINGLE DISH
SOUTH POLE STATION, ANTARCTICA
ELEVATION 9,300 FEET

PDBI
PLATEAU DE BURE INTERFEROMETER
6-ANTENNA INTERFEROMETER
HAUTES-ALPES, FRANCE
ELEVATION 8,400 FEET

IRAM 30M
30-METER TELESCOPE
SINGLE DISH
PICO VELETA, SPAIN
ELEVATION 9,400 FEET

Acronyms and Abbreviations

TELESCOPES AND RELATED TERMINOLOGY

VLBI Very Long Baseline Interferometry, a method in which astronomers observe simultaneously with two or more geographically distant radio telescopes and then combine the data using a supercomputer, simulating a single giant observatory and reaching extraordinarily high levels of resolution.

EHT Event Horizon Telescope, a VLBI array consisting of radio telescopes on several continents, created to study—and image—the black hole at the center of the Milky Way Galaxy and other targets. Over the years, the EHT has included the following telescopes:

> **SMA** Submillimeter Array, a collection of eight six-meter radio antennas at the summit of Mauna Kea in Hawaii.
>
> **SMT** Submillimeter Telescope, part of the Event Horizon Telescope array, located on Mount Graham in Arizona. (Sometimes called SMTO, or Submillimeter Telescope Observatory.)
>
> **CARMA** Combined Array for Research in Millimeter-Wave Astronomy, an interferometer comprising twenty-three antennas located in California's Inyo Mountains. Decommissioned in 2015.
>
> **ALMA** Atacama Large Millimeter/submillimeter Array, an observatory comprising sixty-six movable antennas on the Chajnantor Plateau in northern Chile's Atacama Desert.

APEX Atacama Pathfinder Experiment, a high-altitude telescope in northern Chile.

LMT Large Millimeter Telescope, a fifty-meter single-dish telescope on the summit of Sierra Negra in the Mexican state of Puebla.

SPT South Pole Telescope.

JCMT James Clerk Maxwell Telescope, a single dish on the summit of Mauna Kea in Hawaii.

CSO Caltech Submillimeter Observatory, formerly on the summit of Mauna Kea in Hawaii. Closed in 2015.

IRAM 30M A thirty-meter dish on Pico Veleta in Spain, operated by the Institut de Radioastronomie Millimétrique.

PDBI Plateau de Bure Interferometer, a set of six fifteen-meter antennas installed at an elevation of 8,400 feet in the French Alps.

VLBA Very Long Baseline Array, a permanent network of ten radio telescopes in Hawaii, California, Washington, Arizona, New Mexico (two), Texas, Iowa, New Hampshire, and the U.S. Virgin Islands.

LOFAR Low-Frequency Array, a collection of approximately twenty thousand small radio antennas in the Netherlands designed to gather radiation from the cosmic dark ages.

FPGA Field-Programmable Gate Array, an integrated circuit that a user can reconfigure by writing new code.

LIGO Laser Interferometer Gravitational-Wave Observatory, which detected gravitational waves in 2016, confirming a major prediction of Einstein's theory of general relativity.

CLEAN A set of algorithms developed to create images out of data gathered by radio telescopes.

CHIRP Continuous High-Resolution Image Reconstruction using Patch priors, an imaging algorithm developed by Katie Bouman.

UT Universal Time, a time standard used to coordinate astronomical observations. Universal Time is based on the average speed of

Earth's rotation with respect to the stars, which slows over time. For practical purposes, UT is the same as Greenwich Mean Time.

ASTRONOMICAL OBJECTS

Sgr A* Sagittarius A*, the (suspected) four-million-solar-mass supermassive black hole at the center of the Milky Way Galaxy.

M87 The elliptical galaxy that dominates the Virgo Galaxy Cluster. The term "M87" is often used to refer to the three-and-a-half-billion-solar-mass supermassive black hole thought to reside at the center of that galaxy. Also known as Virgo A.

3C 279 A quasar that the EHT often used as a calibrator source during observations.

3C 273 The first quasar to be identified.

G2 An object, thought to be a giant gas cloud, that astronomers predicted would be torn apart by the black hole at the center of the Milky Way Galaxy.

INSTITUTIONALIA

SAO Smithsonian Astrophysical Observatory, the research institution that manages the Submillimeter Array.

CfA Harvard-Smithsonian Center for Astrophysics, the institution that houses the Smithsonian Astrophysical Observatory.

NSF The National Science Foundation, the United States government agency that supports scientific research outside of medicine.

ASIAA Academia Sinica Institute of Astronomy and Astrophysics, a research institution based in Taiwan. Manages the James Clerk Maxwell Telescope on Mauna Kea.

MSIP Mid-Scale Innovations Program, a funding program for astronomical research run by the National Science Foundation.

Selected Cast of Characters

IN ORDER OF APPEARANCE

AT HAYSTACK OBSERVATORY

Sheperd S. Doeleman, better known as Shep, graduate student and, later, director of the Event Horizon Telescope

Alan Rogers, Haystack Observatory research scientist, radio astronomy pioneer, and thesis advisor to Shep

James (Jim) Moran, Harvard professor and radio astronomy pioneer

Mike Titus, Haystack Observatory correlator engineer

*EARLY PURSUERS OF SAGITTARIUS A**

Donald Lynden-Bell, English astrophysicist who, in the late 1960s, theorized that most, if not all, spiral galaxies contained supermassive black holes at their centers

Ron Ekers, Australian radio astronomer who, with Lynden-Bell, conducted an early search for a black hole at the center of the Milky Way Galaxy

Bruce Balick and Bob Brown, American radio astronomers who, in 1974, made the first detection of what would later be called Sagittarius A*

AUTHORS OF THE 2000 SHADOW PAPER

Heino Falcke, German astrophysicist, then at the Max Planck Institute for Radio Astronomy in Bonn

Fulvio Melia, University of Arizona astrophysicist
Eric Agol, astrophysicist, then at Johns Hopkins University

EARLY EVENT HORIZON TELESCOPE COLLABORATORS

Jonathan Weintroub, South African electrical engineer turned Harvard-Smithsonian Center for Astrophysics astronomer
Avery Broderick, American astrophysicist who performed early computer simulations of Sagittarius A* at Harvard University
Avi Loeb, Harvard University professor and CfA astrophysicist

AT MAUNA KEA (2012)

Rurik Primiani, engineer for the Submillimeter Array, collaborator with the Event Horizon Telescope since 2008
Jonathan Weintroub

BLACKHOLECAM

Heino Falcke
Luciano Rezzolla, Italian astrophysicist who specializes in gravitational waves
Michael Kramer, pulsar expert at the Max Planck Institute for Radio Astronomy

EVENT HORIZON TELESCOPE POSTDOCTORAL RESEARCHERS AND GRADUATE STUDENTS

Laura Vertatschitsch, electrical engineer and radar and field-programmable gate array expert
Michael Johnson, astrophysicist who came to the EHT from the University of California, Santa Barbara
Katie Bouman, MIT computer-vision specialist who developed CHIRP algorithm
Lindy Blackburn, astrophysicist who came to the EHT from LIGO
Andrew Chael, graduate student in the Department of Astronomy at Harvard University

WITH THE ALMA PHASING PROJECT

Mike Hecht, Haystack Observatory assistant director
Geoff Crew, Haystack Observatory research scientist
Lynn Matthews, Haystack Observatory research scientist
Vincent Fish, Haystack Observatory research scientist
Shep Doeleman, principal investigator for the ALMA Phasing Project

AT THE LARGE MILLIMETER TELESCOPE (2014)

Jonathan León-Tavares, Mexican astrophysicist
Arak Olmos Tapia, LMT site manager
Jason SooHoo, Haystack Observatory IT manager
Patrick Owings, hydrogen-maser technician from Microsemi
Gisela Ortiz, graduate student at National Autonomous University of Mexico
Shep Doeleman

THEORETICAL PHYSICISTS STUDYING QUANTUM BLACK HOLES

Steve Giddings, University of California, Santa Barbara, theoretical physicist who proposed that quantum fluctuations at the event horizon of a black hole might be visible to the EHT
Joseph Polchinski, late University of California, Santa Barbara, physicist who led the group that introduced the black hole firewall problem
Stephen Hawking, late University of Cambridge physicist who in 1974 discovered that black holes should destroy information, introducing what has become known as the black hole information paradox

AT THE EHT 2014 MEETING IN WATERLOO, ONTARIO

Remo Tilanus, European project manager on the JCMT
Gopal Narayanan, astronomer at the University of Massachusetts Amherst who built a receiver for the LMT
Dan Marrone, University of Arizona astrophysicist who works on the SPT

Colin Lonsdale, Haystack Observatory director
Geoff Bower
Laura Vertatschitsch
Heino Falcke
Avery Broderick
Shep Doeleman
Jonathan Weintroub
Geoff Crew
Mike Hecht

AT THE SOUTH POLE TELESCOPE

Dan Marrone
Junhan Kim, University of Arizona graduate student

AT THE LARGE MILLIMETER TELESCOPE (2015)

David Sánchez, LMT operator
Aleks Popstefanija, student of Gopal Narayanan
Pete Schloerb, astronomer at the University of Massachusetts Amherst and principal U.S. investigator for the LMT
David Hughes, director of the LMT
Laura Vertatschitsch
Lindy Blackburn
Gopal Narayanan
Jonathan León-Tavares
Shep Doeleman
Gisela Ortiz

THE BLACK HOLE INITIATIVE

Avi Loeb, director
Shep Doeleman, senior faculty
Ramesh Narayan, senior faculty, astrophysicist, and Harvard University professor
Andy Strominger, senior faculty, theoretical physicist, and Harvard University professor

Peter Galison, senior faculty, philosopher, science historian, and Harvard University professor
Shing-Tung Yau, senior faculty, mathematician, and Harvard University professor

DURING THE 2017 EVENT HORIZON TELESCOPE OBSERVATION

IN CAMBRIDGE

Feryal Ozel, University of Arizona professor who studies neutron stars and black holes
Dimitrios Psaltis, project scientist for the EHT
Michael Johnson
Shep Doeleman
Jim Moran
Vincent Fish
Jason SooHoo

AT THE SUBMILLIMETER TELESCOPE

Dan Marrone

AT THE IRAM 30M

Thomas Krichbaum
Heino Falcke

AT THE LARGE MILLIMETER TELESCOPE

Lindy Blackburn
David Sánchez
Gopal Narayanan
Aleks Popstefanija
Kamal Souccar, facilities manager at the LMT

AT THE ATACAMA LARGE MILLIMETER ARRAY

Geoff Crew

ON MAUNA KEA

Jonathan Weintroub (at the Submillimeter Array)
Remo Tilanus (at the James Clerk Maxwell Telescope)

PART ONE

THE VEIL AND THE SHADOW

1

GOLDENDALE, WASHINGTON
FEBRUARY 26, 1979

In his forties, when the experiment started to attract media attention, Shep Doeleman worked up a line of canned biography for reporters—*I was never the type of kid who played with telescopes.* He did, however, have a few early encounters with the cosmic. The first happened on a cold February morning in 1979.

Fifteen thousand people had gathered on a high mound in the golden hills of southern Washington to watch the last total solar eclipse to cast its shadow on the Lower 48 states until 2017. Shep's family had arrived the day before in their Chinook RV camper. The Astronomical League had declared this hill, home to Goldendale Observatory, the official eclipse-watching headquarters of North America. Spectators wore welding goggles, cardboard masks, and strips of Mylar film fashioned into primitive Oakleys. Mothers carried babies against their shoulders, the babies inside protective paper bags. A group of students from the University of Oregon led the crowd in a chant: "E-C-L-I-P-S-E. What does it spell? ECLIPSE!"

Correspondents for the national television networks looked solemnly into their cameras and told New York, *Here we are, minutes from totality, and I'm afraid the weather is not cooperating.*

Alas. The sky was a lead blanket.

Chant leaders urged the watchers to blow the clouds away. The crowd complied with a collective whistling *whoosh.*

The moon slowly began to overtake the sun around 7:15 A.M. The sun pierced the thinning clouds—and then the clouds regrouped, blocking the spectacle. Shep stared hard at the killjoy sky, framing his target with a smoky sheet of Mylar.

The partial eclipse, viewed through a cloud bank, did not exactly inspire awe. But seconds before the moon slid into a place of total obstruction, the plunging temperature dispersed the clouds.

The moon pulled a black tarp across Earth, and just like that, it was night. A yellow ring appeared in the sky. The spectators took off their welding goggles and cardboard masks and strips of Mylar and watched prominences arc though the solar atmosphere. Sunbeams wormed through the moon's valleys and canyons, emerging as beads of light on the shadow's rim.

People screamed, people cheered, and a few set off Roman candles. But after an initial outburst, most stared silently.

Shep had known, intellectually, what an eclipse involved. He was halfway through high school, three years ahead of schedule. He knew the mechanics. Yet he was still unprepared for the full opening-of-the-sixth-seal awfulness of the event. The spectacle seared a high-ranking slot in his memory bank.

People who asked the adult Shep Doeleman if he played with telescopes as a kid were trying to ask a different question: what sort of person makes it his life's work to build an Earth-size telescope to take a picture of a black hole? The eclipse wasn't the cause, but it was an ingredient in the formula. If nothing else, it was his first interaction with the weather gods, those capricious beings that govern access to the skies. They had treated him well. Decades later, he hoped to remain in their graces.

2

In another era, during another eclipse, an astronomer named Arthur Stanley Eddington got his own last-minute reprieve from the weather gods. This one changed history.

It was May 29, 1919, and Eddington was leading an expedition of Britain's Royal Observatory to Principe, an island off the coast of West Africa. At 2:13 P.M. Greenwich Mean Time, the moon was scheduled to blot out the sun, creating an opportunity. In the darkness, stars very near the sun's edge—stars that would normally be washed out by the sun's glow—would become visible. Scientists around the world were profoundly interested in these stars. They offered the best chance yet to test the German physicist Albert Einstein's new theory of relativity.

Eddington got the job because he was an esteemed young scientist, and also because he was in a bit of trouble. He was a Quaker and a pacifist, and, during Britain's war with Einstein's home country, a conscientious objector, but in the final years of fighting his pacifist campaign was not going well. He was about to get sent to potato-peeling camp in Northern Ireland when his boss, Sir Frank Watson Dyson, the Astronomer Royal, came up with a plan.

In March 1917, astronomers at the Royal Observatory realized that the eclipse happening two years hence would be perfect for testing Einstein's new ideas about space, time, and gravitation. During that eclipse, the sun would be surrounded by an unusually large number of bright stars belonging to the Hyades cluster. Einstein's theory predicted that the sun's gravity would bend the light from those stars closest to the sun's edge by twice the amount predicted by Isaac Newton's theory of gravity. They just needed astronomers to travel into the path of the eclipse, take pictures of these stars, and find out which theory was right. It was decided that they'd send two teams. Andrew Crommelin and Charles Davidson would lead an expedition to Sobral, a city in northern Brazil. Eddington and a clockmaker named Edwin Cottingham would travel to Principe.

The mission had all the timeless hallmarks of a big astronomical experiment. Preparation involved committees (Joint Permanent Eclipse Committee) known by acronyms (JPEC); grant applications (100 pounds for instruments, 1,000 pounds for the expedition); the building and borrowing of equipment; multiple layers of uncertainty—while they were getting ready, World War I raged—and, after the fighting stopped on November 11, 1918, a last-minute scramble.

There were long voyages—in Eddington's case, by sea, from Liverpool to Madeira and then on to Principe—and long sojourns far from home. Eddington spent a month marinating in Madeira, taking walks among the eucalyptus and magnolia, playing roulette, hanging out with a dog named Nipper. They dealt with customs agents and called in favors from government officials. They fumbled through foreign languages and attempted to learn the customs of the locals. There were odd, unexpected adventures, like the time the owner of Roça Sundy, the plantation they chose to stage their observation, took Eddington, a pince-nez–wearing Cambridge don, on a monkey hunt. There were days of setting up and testing equipment. Finally, there was an utter dependence on the weather.

In his official report on the expedition, Eddington wrote that between May 10 and May 28, no rain fell on Principe. Naturally, on the morning of the eclipse, it was pouring.

• • •

Historians have cast the 1919 eclipse expedition, in which English astronomers traveled around the world to test a German's theory, as both a gesture of postwar reconciliation and a nationalistic response to a foreign affront to the esteem of Isaac Newton, England's secular god. But you have to suspect that most people involved just wanted to know whether Einstein was right. If he was, that meant the universe was a deeply strange place that people were only beginning to understand.

The principle of relativity, as opposed to the *theory* of relativity, is old and unobjectionable. Bertrand Russell explained it as clearly as anyone ever has. "If you know that one person is twice as rich as another, this fact must appear equally whether you estimate the wealth of both in pounds or dollars or francs or any other currency," he wrote in *The ABC of Relativity*.

> The same sort of thing, in more complicated forms, reappears in physics. Since all motion is relative, you may take any body you like as your standard body of reference, and estimate all other motions with reference to that one. If you are in a train and walking to the dining-car, you naturally, for the moment, treat the train as fixed and estimate your motion in relation to it. But when you think of the journey you are making, you think of the earth as fixed, and say you are moving at the rate of sixty miles an hour. An astronomer who is concerned with the solar system takes the sun as fixed, and regards you as rotating and revolving; in comparison with this motion, that of the train is so slow that it hardly counts. . . . You cannot say that one of these ways of estimating your motion is more correct than another; each is perfectly correct as soon as the reference-body is assigned. . . . And as physics is entirely concerned with relations, it must be possible to express all the laws of physics by referring all motions to any given body as the standard.

This general principle dates at least to the seventeenth century, when Galileo deployed it in argument against those who insisted that Earth couldn't *possibly* rotate on its axis and orbit the sun. If the

Earth is spinning and flying through space, then why, these people asked, presumably with smug, punchable smiles, don't we sense that motion? In his 1632 *Dialogue Concerning the Two Chief World Systems*, Galileo answered with a thought experiment.

Imagine yourself belowdecks on a ship in port. You're in a windowless cabin. Butterflies have made their way into the room. "With the ship standing still," Galileo writes, "observe carefully how the little animals fly with equal speed to all sides of the cabin." Now you set sail. Once you've reached a steady speed, check on the butterflies. Are they "concentrated toward the stern, as if tired out from keeping up with the course of the ship"? Obviously not. That's because the "ship's motion is common to all the things contained in it, and to the air also." This is Galilean relativity. Isaac Newton folded it into his own theory of the solar system, formulated like so: "The motions of bodies included in a given space are the same among themselves, whether that space is at rest or moves uniformly forward in a straight line."

There was no urgent reason to revisit the principle of relativity until the late nineteenth century, when the Scottish physicist James Clerk Maxwell developed his theory of electromagnetism. Among other things, Maxwell's equations predicted that the speed of light through empty space was a universal constant. It never varied, and nothing could travel faster. This presented a direct conflict with Newtonian physics, which had ruled the world since the late seventeenth century. If you shined a flashlight from the prow of a speeding train, Newton's laws said, the light would travel at its usual speed *plus* the speed of the train. Who was right? The biggest minds of the era applied themselves to the conflict. Hendrik Lorentz and Henri Poincaré made progress. But Albert Einstein was the first to find a deep solution.

In his 1905 paper "On the Electrodynamics of Moving Bodies," Einstein proposed a new, two-part principle of relativity. Just as Galileo said, the laws of physics are the same for anyone in an inertial reference frame—someone at rest or in uniform motion. Following Maxwell, he added a major provision: the speed of light is a universal constant. It doesn't matter how fast you're moving, and it doesn't matter how fast the source of light is moving: light always travels through empty space at 186,000 miles per second. Moreover, nothing can move faster than the speed of light. It is the cosmic speed limit.

These postulates lead to counterintuitive conclusions. What if you're in a spaceship traveling at 99 percent of the speed of light? Do you "catch up" with light? Nope: you see light travel at 186,000 miles per second. That's because other things that seem immutable—distance and time among them—are, in fact, flexible. Evolution didn't prepare us for this. Length contraction and time dilation only become noticeable at velocities approaching the speed of light. The fastest thing our brains ever had to process during those formative millennia on the savanna was a sprinting cheetah.

But as an old instructor of Einstein's showed, the strange effects of relativity become natural if you think of the world as spacetime. His name was Hermann Minkowski, and he explained his geometrical interpretation of Einstein's new ideas in a famous lecture in 1908. Space and time were not identical, Minkowski explained, but they were inseparable. "As they occur in our experience places and times are always combined," he said. "No one has observed a place except at a time, nor a time except in a place." Space and time, then, he argued, should be treated as inextricable threads of a unified fabric. To think this way, you have to use a new, unfamiliar form of geometry in which the old Euclidean rules about parallel lines and triangles no longer hold. As a reward for these contortions, you develop an intuitive understanding of time dilation, length contraction, and other relativistic effects.

In Minkowski's spacetime, a thing happening at a time and a place is an *event*. Events are located using four coordinates: three spatial terms (where), and one time term (when). The separation between events—the space-time interval between them—is part distance, part time. Relativity says that the spacetime interval between two events is the same for all observers in all reference frames, regardless of how they're moving relative to one other. Different observers might disagree on time and length, but that's because time and length are not fundamental. As Eddington explained in his book *Space, Time, and Gravitation*, published four years after his expedition to Principe, "*Length* and *duration* are not things in the external world; they are relations of things in the external world to some specified observer."

As mind-expanding as Einstein's new theory of relativity was, it contained a big hole. It set the speed of light as an insurmountable speed barrier. But gravity seemed to travel faster—instantaneously, across the universe. In Newton's world, planets and moons seemed to organize themselves into orbits using magical tractor beams that knew no speed limit. The discrepancy gnawed at Einstein until one day in 1907, as he sat at his desk in the patent office in Bern, writing an article about relativity as it was understood so far, he had the insight he called his "happiest thought." Its significance would have been lost on anyone else: "If a person falls freely he will not feel his own weight."

If falling freely makes you feel weightless, then there is no way to tell the difference between free fall and the absence of gravity. It follows, Einstein reasoned, that the inverse is true: there is no way to tell the difference between accelerated motion and the presence of gravity. When you're pressed into the floor of an elevator that some mechanical glitch yanked suddenly upward, that heaviness you feel is somehow identical to gravity. It took eight years of work to build the general theory of relativity on this insight. It was a difficult theory based on esoteric mathematics, but at the core was a simple, profound idea: Gravity isn't a force at all. It is the curvature of spacetime.

Aristotle ascribed gravity to the self-sorting tendency of all things. Heavy objects "want" to fall toward the center of Earth, and fire yearns toward the heavens. Copernicus thought gravity was "a natural striving with which parts have been endowed . . . so that by assembling in the form of a sphere they may join together in the unity and wholeness." For Newton, gravity was one of the "natural powers" that governed the motion of particles through space. Every particle in the universe attracted every other particle. This force was delivered instantaneously at infinite distances. Newton didn't claim to know *why* gravity worked. Near the end of his magnum opus, the *Philosophiae Naturalis Principia Mathematica*, he wrote, "I have not been able to discover the cause of those properties of gravity from phenomena, and I frame no hypotheses. . . . To us it is enough that gravity really does exist, and act according to the laws which we have explained."

A couple of centuries later, Einstein seemed to have figured it out. Picture the lines of longitude on a globe. Zoom in on a small enough patch, you can ignore the curvature of the globe, and the longitude lines look like they'll never intersect. That's because the patch of globe you're looking at is effectively *flat*—the rules of Euclidean geometry hold, including the one that says parallel lines never intersect. Now zoom out so you're looking at the planet whole. Lines of longitude do intersect, at the North and South Poles. They're still parallel, but they're parallel on a curved two-dimensional surface.

Add two more dimensions to the surface of the globe, one of space and one of time, and you have four-dimensional spacetime, the playing surface of Einstein's general theory of relativity. The mind's eye can't handle curved spacetime. But with training, people can easily write down and manipulate equations that describe how it works. The equations that compose Einstein's general theory of relativity describe the relationship between mass (another form of energy) and the shape of spacetime. As the Princeton physicist John Wheeler wrote, "Spacetime tells matter how to move; matter tells spacetime how to curve." To paraphrase Dimitrios Psaltis, a scientist Shep Doeleman would encounter later in life, gravitating mass causes nearby objects to tilt their futures in its direction. Curved spacetime is not merely a matter of geometry: it's a matter of fate.

• • •

Eddington was too busy feeding fresh photographic plates into his instruments to pay close attention to the sky. "It appears from the results that the cloud must have thinned considerably during the last third of totality," he wrote. These results, combined with those from

Sobral, showed a telltale displacement of the stars, a small discrepancy that heralded the greatest shift in our understanding of space and time since the Enlightenment.

Eddington's results helped make Einstein the most famous scientist of the twentieth century. The general theory of relativity was a scientific triumph and a popular sensation. To the public, Einstein's equations had the holy glow of inscrutable ancient script. "It is as if a wall which separated us from Truth has collapsed," the physicist Hermann Weyl wrote. "Wider expanses and greater depths are now exposed to the searching eye of knowledge, regions of which we had not even a presentiment."

3

What sort of person makes it his life's work to build an Earth-size telescope to take the first picture of a black hole? Someone who's in the right place at the right time, obviously. But only a person with a certain mix of talents, needs, and tendencies—a restless energy, a tolerance for risk and discomfort, and a gnawing need for validation—would, presented with such an opportunity, commit.

Reviewing the biography, it's not hard to see how Shep Doeleman acquired this mix. The restless energy, the seeker's tendency, is probably inherited. He was born in 1967 in Wilsele, Belgium, to young American parents. His father, Allen Nackeman, was twenty-three, and he had come to Europe to pursue medical school. His mother was Lane Koniak, a twenty-one-year-old girl from East New York. The medical school thing didn't last long. When Shep was five months old, Allen and Lane returned to the States, rented a three-quarter-ton truck, and set out for Alaska. They got as far as Portland, Oregon. There Allen found a job as a reporter with the Associated Press, and they settled in the suburbs.

Shep was always a smart kid. Family lore holds that he was reading by three, and that in second grade, after he'd switched from a Montessori school to the local elementary school, he came home one day agitated, holding an insultingly easy spelling test. He thrust it at his mother and demanded something better. Someone told Lane about a private school they should see, a Hillel in the basement of a schul. When they visited, she told the rabbi they had very little money to pay for admission, but the rabbi said they wanted the kid and his little sister, Jeffa, too.

They moved around some, suburb to suburb, but for the next few years, the large-scale wandering stopped. This was apparently too much for Shep's father. When Shep was seven, Allen left for a motorcycle trip across Asia and never came home.

Next came the *Brady Bunch* phase. Lane married Nels Doeleman, a high school science teacher, who brought two kids of his own to the clan, a boy and a girl. Nels adopted Shep and his sister. Years later, Shep would refer to Allen as his "biological father." Nels he called Dad.

After Shep finished fifth grade, the newly configured Doeleman family—two parents, two brothers, two sisters, and a Weimaraner—set out on a big adventure. The parents thought the kids were becoming too Americanized, so Nels took a one-year sabbatical from the high school and they packed the Chinook camper and drove to Montreal, where they boarded the Polish ocean liner *Stefan Batory* and sailed for Belgium.

They spent the year in Louvain-la-Neuve. Shep didn't speak French, but he attended sixth grade in the French-speaking school anyway. Outside school, the family explored Europe on the cheap, driving the Chinook to Italy and Spain and living on big batches of soup and horse-meat sandwiches. It was a good, formative time, but reentry was difficult. That year in an accelerated European school system put Shep ahead of the average Portland seventh grader, so back in Oregon, Shep became a twelve-year-old high school freshman. Both Lane and Nels worked at the high school, so they could keep an eye on him, but the age difference between him and his classmates was hard to surmount. He got picked on, played no sports, didn't go to prom.

But he did take his dad's physics class, and he was a natural. Nels recalls Shep intuitively understanding concepts that he'd struggled with as a student. The science exposure continued at home. There, Nels and Shep launched handmade rockets. In the eastern Oregon desert, they gathered thunder eggs, globules of molten rock with agate crystalline interiors, which Nels would slice in half with a diamond saw in the shed. And one time, of course, they drove a few hours to watch a total solar eclipse.

Shep graduated from high school when he was fifteen. He applied to the California Institute of Technology, or Caltech, Mount Olympus for physics savants, but didn't get in, so he enrolled that fall at Reed College, Portland's temple of freethinking and permissive drug use. He was too young to get a driver's license, so he moved to campus, where, on his mom's instruction, he lied about his age—told everyone he was seventeen. For the next four years he maintained this fiction with such diligence that at graduation, when the speaker announced that his class contained the college's youngest graduate ever, Shep shrank in his seat to hide from his classmates.

One day during his senior year, Shep was walking through the physics department when he found himself transfixed by a notice on the jobs-and-internships corkboard. The Bartol Research Institute at the University of Delaware was looking for two technicians to run scientific experiments for a year in Antarctica. He had already been accepted to the doctoral program in physics at the Massachusetts Institute of Technology—Mount Olympus East—but he could feel the corkboard bending his worldline.

Months later, Shep's bunny boots crunched the groomed-ice runway at McMurdo Station. The C-141 Starlifter that had albatrossed him over from New Zealand whirred, warm from the flight. A red-and-white six-wheeled people mover called Ivan the Terra Bus waited. The sky was ultraviolet. For somewhere between a moment and several days, he thought, *What have I done?* But in time the ice became a part of him.

His lab was a blue metal shoe box called Cosray about a mile outside the main village of McMurdo. He was responsible for running several experiments there, including machines that counted the neutrons knocked out of the atmosphere by incoming cosmic rays. Most of the time Shep would sleep in a dorm at McMurdo and walk or drive the frozen dirt road to work, but Cosray had a little kitchen and a cot, so he could stay in his lab for days, listening to tapes on his Walkman, playing with the Apple Macintosh he had shipped down from Oregon. The lab also had a darkroom, where Shep developed photos he'd taken of the aurora borealis.

The isolation suited him. The screeners from the Antarctic Development Squadron Six, the Puckered Penguins, the naval unit that administered American activities "on the ice," had been worried about this part. In winter, six months of darkness descend and all travel to and from Antarctica stops. Emergency evacuations can only be done at ludicrous cost and grave peril. Research showed that young, unmarried, college-educated people tended to handle Antarctic winters best, and Shep fit that description, but to an extreme. At nineteen years old, he would be the youngest person in the history of the U.S. Antarctic program to winter over.

They didn't need to worry. When the long night arrived, something nocturnal within him emerged. The sky over the ice was a revelation. Stars emerged with impossible density; the longer you stared, the more would appear, darkness slowly filling with points of light in an infinite regress. The boundary between the terrestrial and celestial evaporated. Shep found himself living not on a planet but in a galaxy, not in the present but in the timeless bulk; events thousands of years distant hung immediate in the sky. Later in life, he'd come to think of the universe as a sort of time machine. To look at the cosmos is to look at things that happened thousands or millions or billions of years ago. If the sun vanished, we wouldn't know it for eight minutes. If aliens on the other side of the Milky Way looked at Earth, they'd see Neanderthals.

After he left the ice, he wandered. MIT wouldn't let him defer his application a second time, so he reapplied, and was reaccepted. He knew his wandering years were nearly up, so he biked around New Zealand, then traveled to Asia. In Pakistan, a friend from Antarctica arranged for him to dine at the minister of finance's house. In Kashmir, he rode on top of buses through harrowing mountain passes, peering over the edges of cliffs at the carcasses of fallen vehicles. He communicated with friends and family through exorbitantly expensive phone calls and cellophane-thin aerograms. Finally, in the spring of 1988 he flew to Los Angeles and made his way to New Mexico.

Lane and Nels had split up, and his mom was living in the blasted atomic desert of Alamogordo, teaching at a school for the blind. She took a mother's inventory of her boy's characteristics. She hadn't seen him in nearly two years. Antarctica had not infected him with any obvious psychological maladies. He was still talkative and quick to laugh. Something had changed, though. At minimum, his hair. All his life, Shep had reddish hair. He got it from his biological father, whose beard would come in half red and half brown. Now Shep's hair, still thick with youth, was dark brown, like his mother's.

A few months after he returned to the United States, Shep arrived on the MIT campus. As his ride drove away, he stood on a sidewalk next to a steamer trunk containing the sum of his possessions and realized that lodging—the need to arrange it in advance—had never occurred to him. From a pay phone, he called the professor who had been assigned to be his advisor, a man he'd never met, the head of one of the world's premier research programs into nuclear fusion, and asked if he had any leads on apartments. He did not. And so began Shep's inauspicious graduate school career.

Shep spent much of his time in MIT's Building 37, a rectangular grid of gray concrete beams and narrow black windows. It looked like a giant microchip. Inside it was all bare concrete, fluorescent lights, and discomfort. It could have been designed by an alienation-optimization algorithm.

The same could have been said of the doctoral program in physics. The program rewarded unceasing labor and a high threshold for pain. Naturally, the material was difficult. A single take-home problem set might require thirty pages of calculations to solve. That was okay. Students came to MIT for initiation into an order, to derive the principles of nature from sparse data and fundamental precepts. In an earlier century they might have become schoolmen, or monks. But few of them would have enlisted in the infantry, and at times, that's what it felt like they had done. The program valued hierarchy and the culling of the weak. Professors and advisors would routinely and unironically ask students, *Are you swimming, or sinking?* The words could have been chiseled over the threshold of every building on campus.

Shep felt misfiled among the pure-math prodigies and theoretical-physics obsessives. His interests were cosmic and abstract, but his talents were manual and material. He couldn't always summon the amphetamine concentration that thirty-page tensor-calculus problems required. He needed continual, multidimensional stimulation. He didn't last long in the plasma physics lab. He joined an x-ray astronomy group for a while. He had a good run at a lab growing mammalian nerve cells on integrated circuits, but the professor leading that group took a job at another university, leaving Shep labless. After that, somewhat by default, Shep ended up working in a research group led by a radio astronomer named Bernie Burke. Burke lived a smoking-jacket life of academic glamour, jetting around the world serving on distinguished committees with other distinguished fellows. He left Shep under the guidance of a senior graduate student, and Shep stumbled. He failed his second oral exam twice. On the third try, he passed, but by then, Burke had lost confidence. He wanted Shep out of his group, so he arranged some interviews for him at an MIT observatory north of Boston.

And so one day in 1992, Shep and a few other grad students got in a van and drove north out of Cambridge. An hour outside town, among the colonial homes and kettle ponds, they turned right off a two-lane road and followed a government-issue sign up a long forested drive. Halfway up the hill, domes and dishes came into view. A silver half sphere the size of a cottage sat on top of a large white shed. Farther on, two enormous radar antennas made of metallic mesh faced the sky. One lay on its back, like a bowl placed to catch dripping light. At the crest of the hill stood Haystack Observatory's headquarters, a one-story pedestal of rough gray brick holding a great white Epcot Center sphere.

Shep and his van-mates stepped into the lobby and checked in with the secretary. Outside, the Haystack campus looked like the set of a Cold War–era movie about government scientists secretly communicating with aliens. The local tinfoil hats must have had theories about this place. Inside, though, it felt like a branch office of a regional phone company. These were calm days at Haystack Observatory, at least compared to the beginning.

Haystack descended from MIT's Rad Lab, which in the 1940s perfected radar and, in doing so, helped win World War II. After the war, the Rad Lab closed, but the weaponry that radar was built to detect only got scarier. In the early 1950s, motivated by Russian nukes, MIT and the Department of Defense built Lincoln Laboratory, a sort of open-ended Manhattan Project. Before long, hydrogen bombs and intercontinental ballistic missiles made airdropped fission bombs look quaint. The new doomsday scenarios began with thermonuclear weapons arcing over the North Pole. Lincoln Laboratory scientists responded by inventing new instruments that could track those missiles—like the Millstone Hill Radar down the hill from this lobby, which was completed in 1957, just in time to monitor Sputnik.

In the early 1960s, Lincoln Laboratories built Haystack Observatory, its most advanced radar station yet. The main rationale was to communicate with military satellites twenty-two thousand miles above Earth's surface. From the start, however, the scientists who worked there wanted to do more than prepare for World War III. They wanted to do astronomy. As soon as the 37-meter radar antenna inside Haystack's big Epcot Center sphere was finished, Alan Rogers and others pointed it at neighboring planets and helped nail down the basic specifications of the solar system. In preparation for the Apollo 11 mission, Haystack scientists searched the moon for suitable landing sites, sifting the lunar soil with radar waves, and they assured NASA that the *Eagle* would not, as some feared, sink into a pit of lunar dust, stranding Neil Armstrong and Buzz Aldrin in the Sea of Tranquility.

Beginning in the late 1960s, a group led by Irwin Shapiro of

Lincoln Laboratory conducted a series of observations that constituted the so-called fourth test of Einstein's theory of gravity. With Haystack's radar dish, they bounced radio waves off Venus and Mercury at a time when those signals would pass very close to the sun; the sun's gravity delayed the radar signals' round-trip by two-tenths of a thousandth of a second, just as Einstein's theory predicted.

Radar astronomy of this sort involves bouncing signals off planets and moons. *Radio* astronomy is the practice of capturing long-wavelength light emitted by objects in outer space. For the uninitiated, the words "radio" and "telescope" seem to mix senses. This must be why the makers of the 1997 film *Contact* decided that Jodie Foster, playing a character based on the real-life radio astronomer Jill Tarter, should *listen* to her telescope through headphones. It's fun to imagine a theater full of scientists face-palming their way through that scene. The word "radio" is a truncation of "radiation," the emission of energy. The form of energy astronomers capture is electromagnetic radiation—light. Visible light is electromagnetic radiation with a wavelength of between roughly seven hundred and four hundred nanometers, or billionths of a meter. "Wavelength" refers to the distance between individual light waves. Radio light can have wavelengths of anywhere between one millimeter and many miles. We see visible light because it penetrates Earth's atmosphere and travels easily through the water out of which our primordial ancestors crawled. Evolution has prejudiced us against those bands of the electromagnetic spectrum that our eyes can't detect, but light is light. All of it can be used to see.

By the time Shep stepped into the lobby of Haystack Observatory that day in 1992, it was one of the world's leading incubators for new techniques and technologies in radio astronomy. There, in the quiet hallways under the geodesic sphere, Shep made his rounds, meeting with the directors of various research programs, looking for a lab where he could write his dissertation. After a while, someone brought him to the office of one of Haystack's founding fathers, Alan Rogers. He was a slight, soft-spoken man with white hair and an accent acquired during his youth in Rhodesia. With Bernie Burke,

Alan Whitney, Jim Moran, and others, Rogers made his reputation in the 1960s and 1970s by helping to develop a technique known as Very Long Baseline Interferometry. VLBI, as everyone called it, was a way of coordinating geographically distant radio observatories to simulate a single giant telescope. It was by far the highest-resolution technique in all of astronomy, so it was well suited to studying the very distant and the very small.

Shep knew a little about VLBI, and he was intrigued because, despite the ugly acronym, it held the promise of adventure and romance. He'd heard stories about VLBI people loading atomic clocks onto commercial flights during the worst days of the Cold War and flying them to the Soviet Union, tales of expeditions to alluring places like China's Purple Mountain Observatory and the Svalbard Islands. He was especially interested in programs that used VLBI for geodesy, the monitoring of the fluctuating size and shape of our imperfect planet. For these projects, people had to travel the world installing radio antennas in just the right places, no matter how remote or forbidding. Which sounded pretty great to Shep. The problem, for him, was that most of that far-flung antenna deployment was already finished.

But Rogers had another project he thought Shep might like. He described how he was trying to get VLBI to work at the highest possible frequencies of radio light, the unexplored "sub-millimeter" slice of spectrum between the microwave and the infrared. They were close to getting the technique to work with light of three millimeters in wavelength. After that, the goal was light with a wavelength of one millimeter. Once they mastered one millimeter, they'd conquer the submillimeter realm. It would take years. *The work is difficult,* Rogers explained. They were pushing existing technologies beyond their limits and building new instruments from scratch. The experiments often failed. Fieldwork involves lots of sleepless, chaotic nights at faraway, high-altitude telescopes. *The scientific motivation is quite compelling, though,* Rogers said. *One goal is to look deep into the cores of galaxies to study black holes, including the one at the center of the Milky Way.*

4

As soon as Einstein published the general theory of relativity, his equations began sending strange messages. Karl Schwarzschild was the first to receive them.

Schwarzschild was a prominent German astrophysicist, a colleague and correspondent of Einstein. When war broke out in 1914, he left his post as director of the Astrophysical Observatory at Potsdam and volunteered for the German army, but he kept up his academic work from the battlefield. He read the theory of general relativity in the *Proceedings of the Royal Prussian Academy of Sciences* while posted to the Russian front. "As you see, the war is kindly disposed toward me," he wrote to Einstein on December 22, 1915, "allowing me, despite fierce gunfire at a decidedly terrestrial distance, to take this walk into your land of ideas." On that walk, he derived the first exact solution to Einstein's equations, a feat even Einstein thought might be impossible.

Mathematically, general relativity is a set of equations that, when prodded, explode like trick snakes in a can. They're written in the formidable language of tensors, arcane mathematical objects that plenty of math and physics majors never encounter. An "exact solution" to Einstein's equations is another equation, called a metric, that measures distance in curved spacetime. Different spacetimes have different metrics. The spacetime around a spinning star has a different shape from the spacetime around a star that's not spinning, and so on. Schwarzschild's metric described the most basic possible case: spacetime around a nonrotating spherical mass.

Einstein was so impressed with Schwarzschild's metric that he presented to the Prussian Academy in his stead. It's still the first thing a student of general relativity learns, and it remains useful for modeling the spacetime around stars and planets. But from the beginning there was something ominous about the Schwarzschild metric. It predicted that for any given mass, there was a "critical circumference" where strange things happened. Eddington called it a "magic circle" that no experiment can see inside. The critical circumference depends on mass. For our own sun, whose actual circumference is 4,370,000 kilometers, the critical circumference is eighteen and a half kilometers. If, somehow, our sun's two billion, billion trillion kilograms of mass got crammed into a sphere with a circumference of eighteen and a half kilometers or smaller, light emitted from inside could never escape. In mathematical terms, this critical circumference was a singularity—an undefined point, the equivalent of dividing by zero. Not unreasonably, Schwarzschild and Einstein dismissed the "Schwarzschild singularities" as mathematical artifacts with no relevance for the physical world.

But the equations were trying to say something, and as the years went by, they got louder. In the 1930s, when Subrahmanyan Chandrasekhar worked out what happens to stars when they exhaust their fuel and collapse under their own weight, he found that stars less than 1.4 times the mass of our sun would end up as white dwarfs, objects that crammed the mass of a star into something the size of a small planet. Heavier stars would keep collapsing. A few years later, the astronomer Fritz Zwicky discovered that stars too massive to become white dwarfs would become neutron stars, city-size spheres composed of exotic matter that weighed a billion tons per spoonful. What about even heavier stars? Their fate remained a mystery until 1939, when J. Robert Oppenheimer, a Berkeley physics professor and soon-to-be father of the atomic bomb, and his student Hartland Snyder simulated the death of the heaviest stars in unprecedented detail. They concluded, "When all the thermonuclear sources of energy are exhausted . . . a sufficiently heavy star will collapse," and in all likelihood, "this contraction will continue indefinitely." A collapsing star would shrink to an infinitely small, infinitely dense point—another singularity. If you rode to hell on the surface of that star, the whole cataclysm would be over quickly. But if you watched from a distance, the process would last forever. As the star collapsed, light streaming from its surface would redden, and eventually it would appear frozen in time. Both perspectives—the one in which the star vanished in seconds, and the one in which it shrank for eternity—would be equally "real."

Oppenheimer's peers took this in, shifted in their seats, and said, *Well. Fascinating. Shall we now discuss the German doomsday device?* By 1942, Oppenheimer was working as coordinator of rapid rupture for the Manhattan Project.

After the war, military radar engineers looking for peacetime jobs turned surplus radar dishes into instruments of science and started searching the sky. The gentleman astronomers of the era thought of these newcomers and their "telescopes" as less than legitimate. Nonetheless, as Werner Israel later wrote, the radio astronomers "ushered in the most eventful era in the history of astronomy since the time of Galileo." They did this by locating fountains of radio waves flowing from patches of apparently empty sky and passing the coordinates to optical astronomers, who climbed into the focus cages of their cathedral-size observatories and discovered, to general amazement, that those fountains were coming from dim splotches of light that no one had ever paid much regard. They called these objects radio galaxies. Soon radio astronomers pointed their colleagues to an even stranger species: radio *stars*. These would come to be called "quasars," short for quasi-stellar (starlike) radio sources (things that emit radio waves). Maarten Schmidt, the optical astronomer who discovered the first quasar, made the cover of *Time* magazine. John Bolton, the radio astronomer who gave Schmidt the coordinates, did not.

That quasar, 3C 273, was two to three billion light-years away and a hundred times more luminous than anything else in the sky.

The Americans and the Soviets had H-bombed enough atolls by now that everyone knew nature held apocalyptic forces in reserve. But even nuclear fusion was too weak to explain quasars. 3C 273 shone with the luminosity of four trillion suns—but it couldn't be a galaxy of four trillion stars, because it flickered. Four trillion stars can't flicker in unison. That meant that a quasar must possess some single, relatively small engine capable of converting millions of stars directly and *completely* from matter into energy. In a nuclear-fusion reaction—the process by which two elements fuse to form a third, the process that powers stars and thermonuclear weapons alike—only 1 percent of the mass of those elements gets converted to energy, at best. At a loss, scientists turned to Einstein's theory of gravity.

Gravity is by far the weakest of the four known forces of nature, yet falling is a surprisingly powerful act. As an object falls, it gains kinetic energy. The farther it falls, the more energy it gathers. Einstein's equations showed that matter falling onto unimaginably dense objects would approach the speed of light near the end of their journey. If they hit something on the way down, they would release that energy in an explosion far more powerful than a thermonuclear weapon. The link between quasars and the theoretical energy-liberating power of gravity was fuzzy in the early 1960s, but it was clear which direction scientists in search of an answer needed to go.

Most physicists had ignored general relativity for the previous four decades. It just wasn't all that useful. You didn't even need general relativity to send a spaceship to the moon. Newton's two-hundred-fifty-year-old equations sufficed. But there were a handful of scientists who kept up the work that Oppenheimer left off before the war. Among them was Princeton University's John Wheeler, who spent the 1950s simulating the death of stars in ever-greater detail, and who, by the early 1960s, was convinced that Oppenheimer and Snyder were right: when a star above a certain mass dies, it collapses indefinitely.

Explaining the obscene amounts of energy emitted by quasars was considered urgent enough that physicists called an emergency meeting in Dallas from December 16 to 18, 1963—the First Texas Symposium on Relativistic Astrophysics. When Governor John Connally welcomed the scientists in his opening address, he was wearing a sling from taking a bullet a few weeks earlier in the assassination of President John F. Kennedy. Scientists left the Texas Symposium with a conviction that quasars, general relativity, and gravitational collapse were all somehow connected, but they couldn't quite figure out how. An early suggestion was that quasars were giant "superstars" caught in the act of collapse. But timing was a problem. A star would collapse in a day; quasars had been crackling for millions of years. How could a one-time, one-day event fuel processes that lasted for eons?

In the years after the Texas Symposium, the answer became clear. A new generation of theoretical physicists applying new mathematical tools showed that when a star above a certain mass dies, it becomes something very strange. As John Wheeler described it, the collapsing "becomes dimmer millisecond by millisecond . . . and in less than a second too dark to see. What was once the core of a star is no longer visible. The core like the Cheshire cat fades from view. One leaves behind only its grin, the other, only its gravitational attraction." What remains is, in the words of Werner Israel, "an elemental, self-sustaining gravitational field which has severed all causal connection with the material source that created it, and settled, like a soap bubble, into the simplest configuration consistent with the external constraints." This is what Einstein's equations had been trying to say. The singularities that first appeared in Schwarzschild's notebooks have real, physical meaning. We call them by the name John Wheeler gave them during a lecture in 1967: black holes.

If black holes were just stars that trap their own light, it wouldn't have taken five decades for people to accept them. More than two centuries ago, applying Newtonian principles, the English geologist and rector John Michell posited the existence of "dark stars," which were so dense that they trapped escaping light, and Pierre-Simon Laplace mentioned them in the first edition of his *Exposition du Système du Monde*. The dark stars these two envisioned were solid entities made of normal matter that happened to be unimaginably dense. Light could try to escape, but like an underpowered rocket, it would fail to reach escape velocity and it would fall back to the surface.

Black holes, on the other hand, are made of pure gravity. It can be useful to think of them as processes. The cosmologist Andrew Hamilton compares them to waterfalls. Where gravity is weak, spacetime is a glassy river gliding gently forward. It's easy to disturb a calm river. Drop in a log. The log deforms the surrounding water and disrupts its flow. In this analogy, the flow of the water is time. Canoe past a log and the river will gently steer you toward the disturbance. It tilts your future in its direction. And if you get caught in a waterfall, there's no escape: you're going over.

The defining feature of a black hole is the boundary that appears at Schwarzschild's "critical circumference"—the event horizon. In the words of the physicist David Finkelstein, the first to fully grasp the nature of the boundary, an event horizon is "a perfect unidirectional membrane: causal influences can cross it but only in one direction." It is a place, not a surface. If you were to cross an event horizon (without being vaporized by your fall through the surrounding cosmic hellscape) you would notice nothing. No turbulence. No shimmering diaphanous science-fiction membrane displaying memories from your childhood. "No drama," as physicists say. But you could never return.

Inside the event horizon is a vacuum. Empty space. Einstein's equations say that all the matter from the star that formed this black hole is contained within an infinitely small, infinitely dense singularity at the soap bubble's center. The central singularity is sometimes described as a "knot" of spacetime, but no one knows what happens there. At the center of a black hole, established theories of nature fail.

A black hole that's not spinning and has no electric charge is called a Schwarzschild black hole. It consists of an event horizon concealing an empty interior and that baffling spacetime knot at the center. But everything in space spins, so real black holes must, too. Spinning black holes with no electrical charge are called Kerr black holes, after the New Zealander Roy Kerr, who derived the metric that describes them. Kerr unveiled his metric at the Texas Symposium in 1963, and those who were paying attention were astonished. As Chandrasekhar wrote years later, "In my entire scientific life . . . the most shattering experience has been the realization that an exact solution of Einstein's equations of general relativity, discovered by the New Zealand mathematician Roy Kerr, provides the absolutely exact representation of untold numbers of massive black holes that populate the universe."

How Spin Changes a Black Hole

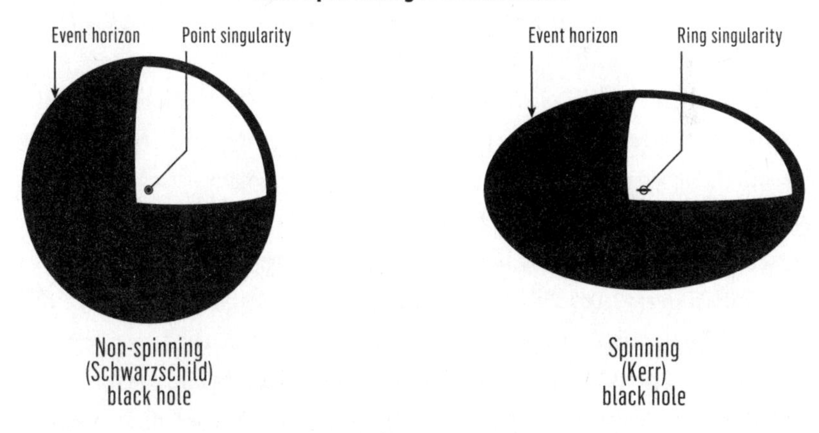

When a star collapses to form a Kerr black hole, the energy that vanished behind the event horizon keeps rotating. Momentum prevents that energy from collapsing into a single, knotlike singularity. Instead, a Kerr black hole has a *ring singularity*, which you can think of as a rotating stream of photons containing the energy of everything that had ever fallen into the black hole. As this slipstream of singularitized energy rotates, it drags spacetime around with it, creating a vortex. In 1969, Roger Penrose used this effect, known as frame-dragging, to show how black holes could power quasars. Black holes do not, despite popular perception, suck in everything around them. But they do eat. Outside a certain distance from the event horizon, matter falls into orbit around a black hole, forming a rotating mass called an accretion disk. As the black hole rotates and drags spacetime around with it, part of the disk gets swept up in the vortex. When matter from the disk enters a region near the event horizon called the ergosphere, Penrose argued, it gets split into two parts. One part falls into the black hole, and the other escapes in the form of energy. The energy that escapes is what makes quasars shine.

• • •

After the first big quasar-fueled burst of work in the 1960s, black hole research settled into two main tracks, one theoretical and one astrophysical. Theorists wanted to know what black holes had to say about the fundamental laws of nature. Astrophysicists and astronomers wanted to find them in the sky.

For a little while, the theorists seemed to have black holes under control. As strange as black holes were, they appeared to be deeply, profoundly simple. A complete physical description of even a speck of dust is more complex than that of a black hole, because that description consists of all the quantum states of every subatomic particle in every atom that makes up that speck. Black holes are different. The so-called no-hair theorem maintains that they can be entirely described by three parameters: mass, angular momentum, and electric charge. They have no bumps or defects, no idiosyncrasies or imperfections—no "hair." Take two black holes. If they have the same mass, angular momentum, and charge, they are *identical*, just as one electron is identical to all others. Throw a thousand-pound refrigerator into a black hole; the hole's mass increases accordingly. Throw a thousand-pound motorcycle into another black hole; the same thing happens. Because those two black holes are identical, all information about the ingredients that went into the black hole—motorcycles, refrigerators, and everything that came before—is hidden behind the event horizon.

This secrecy was all well and good until a young theorist named Stephen Hawking came along. Hawking, of course, was the Einstein of the twentieth century's latter half, famous for his ideas, his wit, and his lifelong battle with motor neuron disease. By the time Hawking finished his Ph.D. at the University of Cambridge in 1966, he was already doing path-clearing research into the objects soon to be known as black holes. With Roger Penrose, he proved that the Big Bang must have begun with a singularity like those at the centers of black holes. He showed that fluctuations in the aftermath of the Big Bang might have created primordial black holes that have been stalking the universe ever since. And with James Bardeen and Brandon Carter, he developed the four laws of black hole mechanics, which were strikingly similar to the laws of thermodynamics. The theorem that a black hole's event horizon can never decrease, for example, sounds a lot like the second law of thermodynamics, which states that the overall entropy, or disorder, of the universe can never decrease.

Most people thought the similarity between black hole mechanics and thermodynamics was nothing but an analogy, but a Princeton graduate student named Jacob D. Bekenstein argued that the area of a black hole *was* its entropy. Hawking hated this idea. If a black hole had entropy, that meant that it had to have a temperature—and everyone knew the temperature of a black hole is absolute zero. It also meant a black hole would have to emit particles,

Black Hole Evaporation

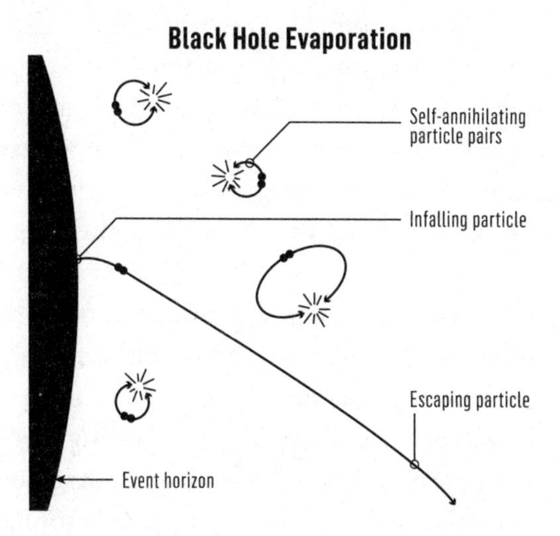

including photons—light particles—and everyone knew black holes are completely black. But in 1973, by then confined to a wheelchair and hardly able to speak, Hawking decided to look at black holes through a quantum-mechanical lens. He quickly reversed his position. In a paper entitled "Black Hole Explosions?" he admitted that Bekenstein was right: black holes have temperature, emit particles, and, eventually, evaporate.

Astrophysically speaking, this revelation was basically meaningless. The temperature of a five-solar-mass black hole would be ten-billionths of a degree Kelvin. It would take many lifetimes of the universe for one to evaporate. But an important principle was at stake. The radiation emitted when a black hole evaporates—Hawking radiation—is random. It contains no information about the stuff that has fallen into the hole. Eventually, a black hole will evaporate entirely. At that point, it would seem that the information about the objects that made up the black hole is lost forever—eliminated from the universe.

Quantum theory famously imposed limitations on our ability to know the world. Subatomic particles with fixed locations turned into clouds of probabilities. Cats in thought-experiment boxes became simultaneously alive and dead. God plays dice, despite Einstein's insistence to the contrary. But while quantum mechanics is probabilistic, it's still deterministic, because it provides definite and unique probability distributions for outcomes. Moreover, quantum mechanics is reversible. In principle, if you could keep track of the quantum changes undergone by every particle in a burning encyclopedia, you could reconstruct the original encyclopedia from the ashes. Destroying data is forbidden. The quantum cosmos remains knowable.

Hawking was saying that if the cosmos contains black holes, it might not be knowable after all. This conundrum, which became known as the black-hole information paradox, revealed that scientists were missing something fundamental about the way the universe works. The paradox wasn't confined to deep space, where actual black holes lived: virtual black holes, like virtual particles, could arise spontaneously out of the vacuum anywhere, so any paradox that black holes pose is therefore universal and immediate. "If determinism breaks down," Hawking explained years later, "we can't know the past. Memories could be illusions. The past tells us who we are. Without it we lose identity." If Hawking was right, either general relativity is wrong, or quantum mechanics is wrong, or information really is destroyed and nothing means anything and physicists should quit their jobs and go work for hedge funds. "Conceivably," the physicist John Preskill later wrote, "the puzzle of black hole evaporation portends a scientific revolution as sweeping as that that led to the formulation of quantum theory in the early 20th century." Theorists would spend the next four decades trying to solve the black hole information paradox.

Astronomers, meanwhile, spent those years hunting for real, live black holes. They catalogued quasars and radio galaxies and other subtypes of active galactic nuclei, the umbrella term applied to all those faraway galaxies that contained furious radiation-spewing reactions at their cores. They launched Geiger counters into space on top of V-2 rockets in search of objects emitting x-rays, and in 1970, they put *Uhuru*, the first x-ray satellite, in orbit. By the mid-1970s, many of them suspected that a source called Cygnus X-1 contained a small black hole feeding like a vampire on an orbiting star. By then, black holes were providing science-fiction writers and prog-rock bands alike with material. Rush eventually released a song cycle about a suicide mission into the black hole at Cygnus X-1.

Through the void
To be destroyed
Or is there something more?

If Cygnus X-1 contained a black hole, it was of the stellar-mass variety, meaning it was formed by the death of a star. These black holes were too small to power quasars and radio galaxies. Quasars were so unimaginably bright that only unimaginably powerful engines could explain them. Those engines, scientists concluded, must be supermassive black holes—those millions to billions of times bigger than stellar-mass black holes.

Most galaxies probably contain supermassive black holes at their cores. Credit for this discovery goes to an Englishman named Donald Lynden-Bell. In 1969, he was the sole theoretical astrophysicist employed by the Royal Greenwich Observatory, which by then had been moved out of light-polluted London and into the countryside. He was thirty-four years old, tall and thin, with a public school pedigree and Cambridge polish. He had been the sort of kid who played with telescopes—a Troughton & Simms three-and-a-half-inch brass refractor that had been in the family since the late nineteenth century. As a boy, Lynden-Bell had trained the instrument on the mountains of the moon and the satellites of Jupiter, but it was other galaxies, wisps of celestial cotton in the refractor's lens, that interested him most. Once he reached adulthood, he spent the first decade of his career trying to figure out how galaxies, particularly spiral galaxies like our own, came together.

Lynden-Bell lived in a village about an hour from the observatory, and most days, he'd carpool with a colleague. They'd drive through the chalk hills and heaths and talk about the genesis of galaxies and the transmutation of elements that took place deep within stars. Along the way, most days, Lynden-Bell noticed a sign for the road A273. It always reminded him of 3C 273, the first quasar.

After seeing that A273 road sign day after day, Lynden-Bell wondered what happened to quasars when they burned out. Quasars were typically billions of light-years away, which meant their heyday was billions of years ago. We don't see quasars nearby. Where did they go? What did they leave behind?

Lynden-Bell began to form an argument. If a quasar is a feeding black hole, then a dead quasar should leave behind a starved black hole. Furthermore, if you take the density of quasars in the early universe and run the statistics for the current, much larger universe, you see that dead quasars should be all around us, cosmically speaking—probably in the nuclei of spiral galaxies such as our own.

In 1970, at Caltech for a long visit, Lynden-Bell and a young Australian named Ron Ekers pointed a pair of radio telescopes at the center of the Milky Way and looked for indirect evidence of a supermassive black hole—gas streaming toward the galactic center at unaccountable velocities. Ekers and Lynden-Bell turned the signals they collected into the first high-resolution radio map of the galactic center. In this map they found odd specks of radiation, and they suspected that one of these specks might be a giant black hole—*the* black hole. But without better telescopes, their argument was weak.

Four years later, two American astronomers, Bruce Balick and Bob Brown, spent a clear February night in the leafless oaky hills of Green Bank, West Virginia, watching the center of the galaxy with better telescopes. Out of caution, they used the phrase "black hole" zero times in the paper announcing their discovery.

Privately, astronomers started talking about "the black hole at the center of our galaxy." In formal publications, they were more cautious. One pair of scientists suggested they call the Balick-Brown object "GCCRS," for "galactic center compact radio source." Bob Brown decided to come up with something better, which couldn't have been hard. The spot resided in a subdivision of sky called Sagittarius A. Brown had studied atomic physics, and in the glyphs of that specialty, an asterisk indicates that an atom is in an excited state. Because the radiation from the (suspected) black hole was "exciting" nearby clouds of hydrogen, causing them to glow, Brown borrowed the asterisk and called the enigmatic spot Sagittarius A*, pronounced "A-star."

5

HAYSTACK OBSERVATORY
1992

Shep took the job in Alan Rogers's lab and bought his first car, a tan 1985 Toyota Tercel. With long-deferred teenage abandon he drove that car to pieces on the expressway between Cambridge and Haystack, where he spent his days studying the doctrine and aims of his new order, this brotherhood of superseers.

For years, Rogers had been developing algorithms, high-speed tape recorders, and advanced signal-processing equipment that could handle submillimeter light. This light was attractive for several reasons. A telescope's resolution—its sharpness, or its ability to distinguish between two objects at a distance—depends on two things: the size of the telescope, and the frequency of the light it collects. Big telescopes collecting high-frequency light have the highest resolution. A big VLBI array collecting extremely high-frequency radio light could reach unimaginable levels of resolution. It could be sharp enough to see into the innermost regions around supermassive black holes.

By 1992, most astronomers were getting comfortable with the idea that Sagittarius A* was a supermassive black hole—our own private quasar. It had taken a while. Bruce Balick and Bob Brown's

1974 discovery of Sagittarius A* became momentous only in retrospect, after other astronomers had gathered supporting data. The first big piece of evidence came in the mid-1980s, when a group led by Charles Townes used infrared telescopes to track gas clouds moving through the galactic center and determined that only the gravitational pull of some very massive, very small object could explain their behavior. Even in 1992, the case wasn't settled. Astronomers had found huge young stars near Sagittarius A*, and the frothing, turbulent space around a black hole should be the least nurturing place imaginable for newborn stars. As a result, some argued that Sagittarius A* *couldn't* be a black hole. Nonetheless, people like Alan Rogers and Shep Doeleman were convinced that if the galactic center wasn't hiding a black hole, it was concealing something even weirder.

Assuming Sagittarius A* was a black hole, there were bigger ones, and closer ones, too. Hundreds of millions of small black holes careen around the galaxy like wrecking balls. But most of these are unobservable, orbs of darkness drifting in darkness. Supermassive black holes were visible because they ate, and so they glowed. After Sagittarius A*, though, the nearest supermassive black holes are in the centers of other galaxies. Sagittarius A*'s enormous size and proximity to Earth made it the biggest apparent black hole in the sky. But a many-layered veil hid Sagittarius A* from human view. Alan Rogers's goal—and, now, Shep's—was to pierce this veil with high-frequency VLBI.

The veil's first layer was the galactic plane, a sheet of gas and dust and dead-star ash lying between our planet and the center of the Milky Way. This fog of interstellar gunk extinguishes the amount of light reaching us from the galactic center by a factor of one hundred million. Only radio waves, x-rays, gamma rays, and a few slivers of near-infrared light make it through. If our planet happened to be above the galactic plane, looking down at the center of the Milky Way like a mountaineer surveying a valley, the glow of the galactic center would overwhelm the sky. Instead, even on the clearest night in the darkest place on Earth, to the naked eye the Milky

Way looks smudged out, like a declassified document with the good stuff redacted. Fortunately for them, radio astronomers didn't have to worry about the galactic plane: radio waves travel through it as easily as cellphone signals sail through drywall.

The veil's next layer, however, was a problem. Astronomers called this layer the scattering screen. They didn't know exactly what it was. Maybe an exploding star had sent shock waves through clouds of dust near the galactic center, setting them aswirl, like freshly poured cream whirlpooling through a cup of coffee. If so, when light from Sagittarius A* flew through this churn, some of it got knocked off course, blurring the view from the other side. Whatever the cause, astronomers could predict how strongly the scattering screen would obscure light of different frequencies. The effect was exponentially stronger for lower frequencies. Nothing disrupts visible light in quite the same way, but for an analogy, picture two rubber ducks of equal size, one red and one blue, behind a frosted shower door. Red light is lower in frequency than blue light. If that shower door behaved like the scattering screen concealing Sagittarius A*, the red rubber duck would look like a smeared-out blob ten times bigger than a blue rubber duck of the same size. Years of study suggested that at the highest frequencies, the scattering screen would become transparent.

High-frequency radio light might also pierce the veil's third layer: the hot atmosphere surrounding the black hole. This matter is the only reason the black hole shines. But the outer layers obscure the fireworks at the core. Seeing through this obstacle would be a matter of peeling back layers. Low-frequency radio light comes from the outer layers; higher frequencies originate closer to the center. Everything astronomers knew so far suggested that the highest-frequency radio light—the microwave band—came from the edge of the black hole itself.

The question was whether the atmosphere in these inner layers was opaque or transparent. If it was opaque, astronomers could look as close to the event horizon as they wanted, but all they'd see was a ball of fire. If the atmosphere became transparent, however, they might be able to see all the way in. And who knew what they'd find there.

• • •

The people who invented Very Long Baseline Interferometry didn't know it at the time, but they developed the technique to look at black holes. They had wanted to learn what forces churned in the hearts of quasars, and to do that, they needed the biggest telescopes they could imagine.

Astronomers divide the sky into angular units—degrees, arcminutes, arcseconds. The units work like the intervals on a clock: there are sixty arcseconds in an arcminute, and so on. The most acute pair of human eyeballs can resolve objects roughly twelve inches apart at three-fifths of a mile. In angular units, that is a little more than one arcminute. The best optical telescopes have a resolution limit of about one arcsecond, which is about two kilometers at the distance of the moon. The first radio telescope had a resolution limit of thirty *degrees*, which is the equivalent of sweeping your hand across the night sky and saying, "The radio waves are coming from up there."

That telescope was the invention of Karl Guthe Jansky, an experimental physicist working for Bell Laboratories in Holmdel, New Jersey. It was 1930, and his assignment was to find the source of noise in transatlantic phone calls, which at the time were beamed across the ocean as radio waves. If he could figure out which direction the noise was coming from, Bell engineers could point their antennas away from the source. And so in a field in Holmdel, Jansky built an instrument that looked like a one hundred-foot-long length of half-built scaffolding. The whole thing rotated in place on four Model T wheels. Jansky began studying the sky with this merry-go-round in late 1930, and by January 1932, he knew he was onto a mystery: "a very steady continuous interference—the term 'static' doesn't quite fit it," he wrote, that "goes all around the compass in 24 hours."

For the rest of the year, he recorded and pondered this interference. As the months went by, the Nazis took control of the Reichstag, famine broke out in the Soviet Union, and the first storms of the Dust Bowl began. Farmers revolted, workers struck and rioted. Not far from Jansky's telescope, the son of Charles Lindbergh was kidnapped and found dead weeks later. All in all, not a great year. But good things were happening in Jansky's professional life. By December, he had clarified the problem. By watching the not-quite-static hiss change with the seasons, by patiently eliminating reasonable explanations for the noise—by noticing, for example, that the partial solar eclipse of August 31 had no effect on the hiss, which meant the sun was not the cause—he found that some of the disturbance came from nearby and distant thunderstorms. But he also discovered that the sky was crackling with radio waves issuing from the dead center of the Milky Way. He called it "star noise."

The *New York Times* covered Jansky's discovery on May 5, 1933. "New Radio Waves Traced to the Centre of the Milky Way," read the headline. No one knew what to make of this. Everyone assumed that the sun would be the most prodigious source of radio waves in the sky. Instead, Jansky had found a geyser of invisible light emanating from the core of the galaxy. "There is no indication of any kind," the *Times* assured, ". . . that these galactic radio waves constitute some kind of interstellar signaling, or that they are the result of some form of intelligence striving for intra-galactic communication."

Jansky didn't find Sagittarius A*. He found the whole dense mass of radiating insanity that fills the inner portions of the galaxy. The discovery was monumental, though—and yet the astronomers of the day had little use for Jansky's discovery, because the clues his crude "telescope" had gathered were too vague for them to pursue.

It took an amateur from Wheaton, Illinois, to pick up Jansky's work. In the late 1930s, Grote Reber, archetypal American tinkerer and ham radio enthusiast, taught himself optics at the public library and diverted money a normal young bachelor might have spent on a car to build the world's first parabolic radio telescope in his mother's backyard. Parabolic radio telescopes are the big iconic dishes, and because they collect and concentrate radio light better than the bare antennas Jansky used, their introduction was a major step forward. Reber's dish was more like a farm implement than a modern radio telescope. It was thirty-one feet in diameter and consisted of a wooden frame strung with chicken wire. Still, it was much more accurate than Jansky's antenna. Reber used it to make contour maps of radio emission from the sun, the Milky Way, and powerful, distant sources in the constellations Cygnus and Cassiopeia, fountains of radio signals that a few decades later scientists would come to know as quasars.

As soon as people got serious about using radio telescopes to study the sky, they realized they needed impractically huge dishes to do anything useful. The lower the frequency of light to be observed, the bigger a telescope must be to pick out objects close together on the sky. Because radio light is millions to billions of times lower in frequency than visible light, radio telescopes must be much, much bigger than optical telescopes. That's why astronomers turned to interferometry.

To build a radio interferometer, astronomers take two or more radio telescopes, observe with them simultaneously, and combine the data they collect. Together, these telescopes can achieve the sharpness of a single dish with a diameter the distance between the two telescopes. The technique is called interferometry because when the radio waves from the different telescopes are combined, they *interfere* with one another. If the waves are in phase—if the crests and the troughs line up—they yield a bigger wave. That's positive interference. If the waves are out of phase, dissonance reduces them—negative interference. Positive interference amplifies signal, negative cancels noise.

Australian astronomers built the first radio interferometers in the late 1940s by placing antennas on sea cliffs and collecting radio waves as they bounced off the ocean. A few years later, astronomers at the University of Cambridge, led by Martin Ryle, wired together antennas separated by a few hundred meters. The distance between the antennas was called a baseline. The longer the baseline, the higher the resolution. When the baselines got so long that wires became impractical, astronomers tried beaming microwaves between telescopes. It worked, but only up to a couple hundred miles.

By the early 1960s, scientists at Caltech had built a radio interferometer with five arcseconds of resolution—four-hundredths of the diameter of the sun as seen from Earth. It was an improvement, but nowhere near what it would take to study the cores of quasars. That job called for a radio interferometer that spanned continents. And the only way to build an instrument like that was to record data collected at every site, ship those recordings to a central location, play them back at precisely the same rate, and record a new, summed signal, mimicking the output of an actual intercontinental telescope.

Soviet scientists proposed doing just that in the early 1960s, but to record the data accurately enough at each telescope to add them together later, they needed high-speed tape recorders and ultra-accurate clocks that didn't yet exist. That technology arrived first in the West, in the mid-1960s, and as soon as it did, Alan Rogers, Bernie Burke, and others at Haystack raced two other North American groups to perform the first tape-recorder interferometer experiment. All succeeded by 1967. By 1969, astronomers had pulled off an observation on a baseline longer than ten thousand kilometers between Sweden and Australia, about as far as Earth's geometry will allow. The name they'd been using for this new method—"long baseline interferometry"—was no longer superlative enough. That's when the technique got the name that stuck: *very* long baseline interferometry.

Pretty soon, astronomers realized they could turn VLBI back on itself and study Earth. When two telescopes are thousands of miles apart, light from the same celestial object will reach those telescopes at slightly different times. Measuring that difference makes it possible to compute the distance between those telescopes down to the millimeter. And because quasars are so far away from Earth that, from our perspective, they basically never move, they can serve as fixed reference points on the sky. Astronomers could use the same pair of telescopes to look at the same quasar dozens of times over the course of a few years; if the time delay between the telescopes changed over time, they knew it wasn't because the quasar was moving: it was because the telescopes were moving. Which meant that the *continents* were moving. And that is how VLBI clinched the case for the theory of plate tectonics. "There is a curious parallel between the histories of black holes and continental drift," the physicist Werner Israel wrote. "Evidence for both was already nonignorable by 1916, but both ideas were stopped in their tracks for half a century by a resistance bordering on the irrational." A strange astronomical technique with a clunky acronym forced scientists to accept both.

• • •

Shep's training started in Haystack's correlator room, a big windowless space with a suspended ceiling and fluorescent lights, where a purpose-built supercomputer, one of maybe three such machines on the planet, synthesized the signals collected at different telescopes. The correlator itself was a row of refrigerator-size cabinets lining one wall. Some of these cabinets displayed tire-size reels of magnetic tape mounted on playback heads. Correlator engineers would spin these reel-to-reels like DJs, playing tapes together and recording a new, synchronized master cut. With the right input the correlator would smooth out discrepancies between the tapes, anticipating the nanoseconds-long delay in arrival time from station to station, compensating for subtle, latitude-dependent differences in the wobble and rotation of Earth, erasing misleading Doppler shifts caused by the speed at which a given observatory is rotating toward or away from the source of light, depending, of course, on the spin of the planet itself.

In this room Shep began to learn the art and trickery of VLBI. On those data recorders and playback heads he practiced threading magnetic tape through rat mazes of pulleys and spring-loaded capstans. Mike Titus, who ran the correlator room, and the other veteran tape handlers taught him tricks. When it was time to spool tape from one reel to another, if you turned the spindle just right, loading the takeup reel with static electricity, you could grab the film with your free hand, snake it around the catch, start the reels rolling, let the static electricity on the empty reel grab the end of the tape, and then *slam* the doors on the vacuum chamber; the machine would power on, the vacuum would pull the tape taut, and the reels would start spinning.

The spring after Shep started, he got his first taste of fieldwork. During an experimental observing run, he was posted to the Five College Radio Astronomy Observatory, about seventy miles southeast of Haystack on a forested peninsula jutting into the Quabbin Reservoir. Plans for the reservoir, which the state of Massachusetts built starting in 1936, damming the Swift River, inspired H. P. Lovecraft's short story "The Colour Out of Space," in which a Boston surveyor visits a remote valley west of the fictional town of Arkham and, through conversations with locals, reconstructs a tale of cataclysm from the sky. A meteorite hit the valley, the story goes, delivering a cosmic mutagen that rendered the sugar maple terrain a "blasted heath," infecting the valley's crops and livestock with interstellar leprosy, driving the residents insane and sometimes to their deaths. The agent of destruction was some alien otherness whose distinguishing attribute was a color "almost impossible to describe; and it was only by analogy that they called it colour at all." Which, as it turns out, is not a bad way to think about the color of the radio light Shep had come to this reservoir to collect.

The observatory was isolated enough that there was no easy coming and going. On observing nights, Shep stayed up past dawn, switching huge reels of magnetic tape every two hours. Each day he would kneel next to the hydrogen-maser atomic clock, a metal box the size of a central-air unit, and reach into a small opening with a dentist's mirror, looking for the pink-violet glow of the oscillating hydrogen atoms that gave the clock its unshakable metronome beat. Only if the atomic clocks used at all sites were perfectly accurate, perfectly synchronized, would it be possible to combine signals from the telescopes later.

The job involved days of Waldenesque boredom punctuated by moments of pulse-spiking stress: say, splicing together tape that snapped while the telescopes slewed from quasar to quasar on a tight, fixed schedule. The work suited him. It was hard intellectual and manual labor in pursuit of distant and exotic phenomena, with the promise of interesting travel and purifying isolation.

A year later, Shep made first contact with Sagittarius A*. His mentors had planned a campaign to observe the galactic center at the wavelength of three millimeters, another step toward the submillimeter threshold. The observation set into oscillation the rhythm that would pace Shep's next two and a half decades. There are only a few weeks per year when Sagittarius A* is visible in the Northern Hemisphere and the weather has a chance of being good, and they fall in late March, early April. For Shep, and all the others who would join his quest, early spring became a time of pilgrimage. This spring, after they'd trucked the tapes back to the Haystack observatory correlator room, where Mike Titus synthesized the signals they'd gathered, they could say with new confidence that the gravitating core of Sagittarius A* was so small that it had to be a black hole.

6

HARVARD SQUARE, CAMBRIDGE, MASSACHUSETTS
JUNE 8, 1995

The whole family came to Boston for graduation—even Shep's biological father, Allen. On commencement day, Shep was leading this atomized bunch on a walk through Harvard Square when they stopped by a bookstore to pick up his doctoral hood. Shep, his mother, and his grandmother, Nana, were milling around in the store when Shep stopped thinking about the dynamics of the day. Instead, he was pretending to look at books while looking at a girl. They were standing on opposite sides of a low table stacked with new paperbacks or staff picks or something like that; it didn't really matter, because the books had ceased to be anything but navigational aids in their mutual approach.

"So," Shep said to her. "What book should I get?"

Her name was Elissa, accent on the second syllable, "i" pronounced "ee," Weitzman. She had light skin and wore her dark hair in an explosion of tight curls. She was looking for a gift for a cousin who was graduating from Harvard College. She had the narrowed eyes of the perpetually unimpressed, but she thought Shep was cute. She even found his hapless opening line charming.

They hadn't gotten to the Cambridge ritual of asking each other which school they were associated with when an older woman with a Brooklyn accent grabbed Shep's shoulder and pulled him toward the door. "Shep, let's *go*, we're gonna be late for your ceremony!" Shep's mom had a better sense for these things than her mother, so she scolded Nana in a whisper loud enough to embarrass everyone in the store. "Mom, can't you see what's happening here? Leave him be." She bought them just enough time to exchange phone numbers.

Elissa's signature memory from the earliest days of their relationship is hysterical, pointless laughter. They'd laugh so hard the laughter would amplify itself and they'd forget what they were laughing about. They had a lot in common. They were both more-or-less nonpracticing Jews, and they had both had weird, accelerated childhoods. Elissa grew up in upstate New York, near Utica, the youngest daughter of middle-class parents whose forebears had come to America from Russia and Lithuania a few generations earlier. When she finished high school two years early, her guidance counselor wasn't sure what she should do, but he was *pretty sure* she shouldn't go applying to any Ivy League schools, so she joined an international program in Jerusalem. When she got home, she filled out an early-decision application to Brandeis by hand, in green ink. A friend's older sister had gone there. It was the only school she applied to.

Like Shep, Elissa was a young scientist, but her work could not have been less cosmic. She was an epidemiological fieldworker. She got into it in the early 1990s, doing public-health work for USAID in the slums of Haiti and Bolivia. Back in the States, she went to work for the Boston public policy research firm Abt Associates and traveled to deprived communities around the country, counseling drug users and sex workers on condoms and HIV tests.

Within a few months, Shep and Elissa were making career choices based on mutual mobility. He was offered a position in Japan, but Elissa couldn't go—she had started a doctoral program at the Harvard School of Public Health—so Shep took a one-year position at Haystack that was in no way guaranteed to turn into something permanent. After the year was up, he'd have to apply to keep his own job. But he took to the high-frequency VLBI mission with such zeal that his one-year gig became a three-year postdoctoral position, and before long, he and Elissa moved into an apartment in Cambridge together.

When his postdoctoral fellowship was up in 1998, Shep took a permanent job as a full-time research scientist at Haystack. In a torch pass from Alan Rogers, Shep became responsible for running Haystack's high-frequency VLBI research. The same year, he and Elissa were married. His years of wandering and lab-hopping were finished. And around the same time, people in Shep's loose scientific circle were becoming aware of an opportunity that Shep was well placed to seize.

TUCSON, ARIZONA
SEPTEMBER 7, 1998

The Sheraton El Conquistador was a golf resort in the usual southwestern mode: stucco walls, red-tile roofs, buzz-cut lawns of unlikely green. Sixty-one scientists had gathered here under cobalt-blue skies for a weeklong conference on the innermost sector of the galactic center.

When astronomers gather to share their latest glimpses of the unknown, a hierarchy develops. At the top are those with the most recent hits. The stars of the 1998 Central Parsecs conference were two competing groups of astronomers, one American and one German, who specialized in the collection of infrared light. This band of radiation, just a little too long in wavelength for the unassisted human eye to see, could, like radio waves, make it through the wall of dust blocking our view of the galactic center. Since the early 1990s, astronomers had been using this light to track the motion of stars near Sagittarius A*. And this morning, they brought news of great wonders.

Reinhard Genzel and Andreas Eckart of the Max Planck Institute for Extraterrestrial Physics led the German group. Andrea

Ghez, a thirty-three-year-old professor at the University of California, Los Angeles, represented the Americans. Both groups were using new, high-altitude, high-precision optical telescopes equipped with cameras tuned to infrared frequencies. Both telescopes had flexible mirrors that computer-controlled actuators adjusted thousands of times each minute to focus on dim sources that would otherwise be blurred into extinction by fluctuations in Earth's atmosphere. So equipped, in 1992 Genzel and Eckart started taking annual portraits of dozens of stars within light-days of Sagittarius A*. After a few years, they saw that these stars were moving around some tiny, dark central mass faster than a thousand kilometers per second. That was ten times faster than stars a little farther out from the galactic center.

This morning, the groups described these discoveries, explained how giant blue stars were boomeranging around Sagittarius A* at millions of miles per hour, like planets orbiting an absent sun. It was the best evidence yet that Sagittarius A* could be nothing but a black hole.

After lunch, it was the VLBI crowd's turn. They were on the B-list. People at Haystack and elsewhere were making progress toward the one-millimeter mark, but it was a slog, and they had no splashy discoveries to show off. They continued to peck at the veil surrounding Sagittarius A*, trying to see through the interstellar scattering, hoping that with the right telescopes operating at the right wavelengths the black hole at the center of the Milky Way would reveal itself. Shep and the German radio astronomer Thomas Krichbaum were preparing for a big one-millimeter observing run the following year. They'd pair an observatory a few hours from Tucson, on Mount Graham, with telescopes in Europe, and they'd see if they could at least detect Sagittarius A*. But that was development work, and if they were being realistic, most of the people in the El Conquistador meeting room that afternoon would have admitted that they probably needed a new generation of observatories before they could make a real breakthrough. New telescopes were being built in Hawaii, California, Mexico, and Chile, but these were years-long, multimillion-dollar projects, and most were running behind schedule.

After Shep and others presented their latest vague, conflicting results, discussion turned, as always, to those new telescopes. Thomas Krichbaum told the room that some combination of the new instruments, united by VLBI, should be able to see all the way down to the event horizon of Sagittarius A*. "That means in principle," Krichbaum said, "in a few years' time it should be possible to image an area around the black hole, or as close to the black hole as possible."

After some cross talk about telescope upgrades, a German theorist named Heino Falcke spoke up. He had soft features and spoke in quick, precise English. "We are approaching—really—the black hole at these high frequencies," he said.

Another astronomer asked Heino to clarify.

"At higher resolution," Heino replied, "there literally will be a 'black hole' in the emission region we could observe."

It was not a given that a black hole would literally look like a black hole. Sure, by the late 1990s illustrators had produced enough speculative artist's renderings of black holes to fill a large, mediocre museum. But no one had spent much time building scientifically accurate models of the optical appearance of real black holes, because the chance of ever seeing one seemed impossibly remote.

James Bardeen was an exception. In 1973, he figured out that, theoretically, in the right circumstances—if, say, a black hole passed in front of a large, bright background, like a star—an observer could see its silhouette. At the time, he was a young Yale physicist trying to escape the supermassive shadow of his father, John, the only two-time winner of the Nobel Prize in Physics. For theoretical reasons, the younger Bardeen was working through equations that described how light should move near a black hole. He found that if a black hole happened to pass in front of (say) a giant star, an onlooker would see a black circle glide across the surface of the star. "Unfortunately," Bardeen concluded, "there seems to be no hope of observing this effect."

Later that decade, the French physicist Jean-Pierre Luminet did some similar calculations. Instead of thinking about a black hole passing in front of an external light source, he asked what a black hole would look like if illuminated by the glow from its own accretion disk. The difference between a black hole and any normal object is that a black hole doesn't have a surface that reflects light rays. "It is the black hole's *gravitational field* which deviates the light rays," Luminet wrote. "The trajectories of light rays are not altered by the impact with a surface, but are curved by the gravitational field." The strength of the gravitational field, Luminet found, creates bizarre, funhouse-mirror effects. Picture a familiar photo of the rings of Saturn. The rings circle the entire planet, but you only see part of them—the part that passes in front. The disk around a black hole would look completely different, Luminet learned. "The first surprise is that *all of the upper side of the ring in front and behind the black hole is visible*, including the portion which would 'normally' be hidden. . . ." he wrote. "Even more surprising is that the curvature of space-time around the black hole enables us to observe the *lower side* of the rings. This is the *secondary* image. So it is possible to observe both the upper and lower sides of the accretion disk!" The image is gravitationally lensed—magnified. And there are, in fact, "an infinity of images, because the light rays emitted by the disk can travel any number of times around the black hole before escaping from its gravitational field and being observed by a distant astronomer." Luminet did his calculations by feeding punch cards into a computer, and then he drew the results by hand. His black-and-white images looked like twisted depictions of a black Saturn, with a ringlike accretion disk warped like taffy. Luminet later translated a nineteenth-century French poem by Gérard de Nerval, "Le Christ aux Oliviers," that he imagined described the experience of staring into a black hole.

> *In seeking the eye of God, I saw nought but an orbit*
> *Vast, black, and bottomless, from which the night which there lives*
> *Shines on the world and continually thickens*

A strange rainbow surrounds this somber well,
Threshold of the ancient chaos whose offspring is shadow,
A spiral engulfing Worlds and Days

Heino Falcke didn't know about Luminet's work, but he had come across Bardeen's calculations in the early 1990s as a graduate student at the University of Bonn, and they had lodged in his mind. Throughout the 1990s, as radio astronomers worked to pierce the veil concealing Sagittarius A*, Heino deployed Bardeen's equations in papers and talks to show what they might eventually find there.

The year after the conference in Tucson, Heino was working as a visiting professor at the University of Arizona, collaborating with Fulvio Melia, who specialized in Sagittarius A*. In the mid-1990s, Melia and a student named Jack Hollywood performed their own simulations on the appearance of black holes. They were just exercises. Enough had changed since then that Heino, Fulvio, and the physicist Eric Agol decided to see whether there was a realistic chance of seeing the type of literal black hole Bardeen predicted in Sagittarius A*.

Eric Agol had developed computer software called relativistic ray-tracing code, which predicts how light travels when subjected to gravitational lensing, light bending, frame-dragging, and other extreme relativistic effects. With this code they simulated how Sagittarius A* would appear to a planet-size high-frequency VLBI array. The software predicted that the telescope would see a boundary identical in shape to but ten times larger than the event horizon. At its edge, light rays would be trapped in a perfect circle, tracing a glowing ring. Inside this ring, darkness. Sagittarius A* should cast a "shadow"—a literal black hole. The size of the shadow depended on the mass of the black hole, which astronomers were continually revising as better measurements came in, but roughly, it would be about fifty million kilometers in diameter. To us here on Earth, it would be like looking at a doughnut on the moon. That happened to be just within range for an Earth-size VLBI array collecting light of about one millimeter in wavelength.

The Black Hole Shadow

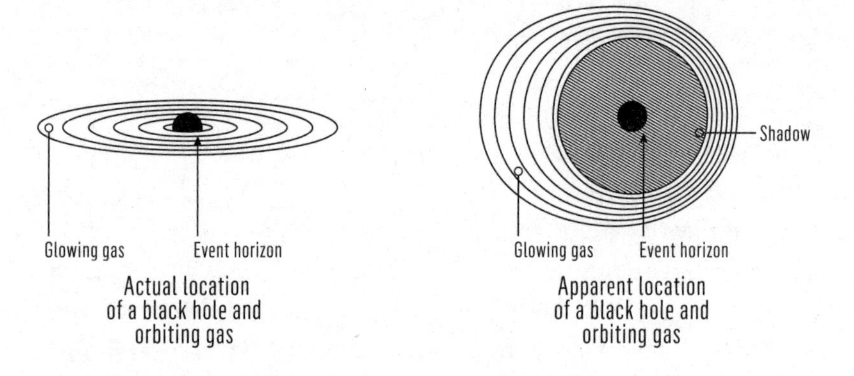

That this shadow might be visible from Earth depended on an astonishing set of circumstances. First, Earth's atmosphere happened to be transparent to the light shining from the edge of the black hole, even though it blocked light of slightly longer and shorter wavelengths. Next, the scattering screen also became transparent at those frequencies. Finally, the atmosphere surrounding the black hole also (probably) became transparent *at those same frequencies.* Later in life, Fulvio Melia compared this alignment to the cosmic accidents that give us total solar eclipses. The moon is just the right size, in just the right orbit, at just the right distance from Earth that now and then it blocks the sun entirely. As if to emphasize these unexpected connections, the black-hole shadow that Heino and Fulvio and Eric Agol predicted looks a lot like a total eclipse of the sun. Fulvio wasn't religious, but these coincidences were so unlikely that he couldn't help but feel that the black-hole shadow was meant to be seen. The universe had arranged for humans to see to the nearest exit.

The Falcke-Melia-Agol trio published their findings in the January 1, 2000, issue of *Astrophysical Journal Letters*, and the Max Planck Institute celebrated with a press release. "First Image of a Black Hole's 'Shadow' May Be Possible Soon," it declared. "Slight advances," it said, could make it happen "in the next few years."

7

Slight advances." "Next few years." These were the words of an overexcited publicist. Nonetheless, in the scientific literature from the first years of the new millennium, you can see signs of astronomers cautiously circling this new prey. *Science* published an enthusiastic article on January 7, 2000, describing how astronomers were closing in on the black hole at the center of the Milky Way. By 2002, Shep was showing diagrams in papers and talks depicting how an Earth-size array of radio telescopes could come together to take Sagittarius A*'s picture. That year, he and Thomas Krichbaum got telescopes in Arizona and Spain to work together at a wavelength of two millimeters—a big step toward the long-sought one-millimeter mark. CNN.com picked up the news, with a headline that read, "New Telescope Is as Big as Earth Itself."

On a Friday evening in late March 2004, at a conference in Green Bank, West Virginia, celebrating the thirtieth anniversary of the discovery of Sagittarius A*, Shep, Heino, and Geoff Bower, a young astronomer from the University of California, Berkeley, made the case for mounting a major effort to look for the black hole's shadow. First, Geoff cued up a PowerPoint deck titled "Road Map to the Event Horizon." He was buoyant from a recent success, the biggest of his career so far: using the Very Long Baseline Array, a permanent network of radio telescopes spread across the United States, he and a group of collaborators had seen through the scattering screen blocking Sagittarius A* at a wavelength of seven millimeters—not close enough to the submillimeter realm to reach the black hole's edge, but getting there. The way he figured it, by next year, they could link up telescopes in Hawaii, Arizona, and Chile, and then each year after that they would add more. By 2009, they'd have the first picture of a black hole in hand. Next, Heino explained why they should go to all this trouble. They could subject Einstein's general theory of relativity to new, exotic tests, of course, but they could also explore some radical, even fringy ideas.

Not everyone thought Sagittarius A* was a black hole. Not everyone thought there *were* black holes. Some people thought Sagittarius A* was a hypothetical creature called a supermassive boson star—something between a neutron star and a black hole, prevented from collapse by the rules of quantum theory, with no solid surface but also no event horizon or central singularity. They weren't exactly sure what a supermassive boson star would look like, but if Sagittarius A* was, in fact, something other than a black hole, they'd know it when they saw it.

After Heino finished, Shep got up and talked about the work this would take, which was significant. Radio astronomers sometimes emphasize the difficulty of their jobs with the following factoid: all the combined photons collected by every radio telescope ever built, excluding those emitted by our own sun, would carry too little energy to melt a snowflake. To compensate for this scarcity—to collect as many photons as possible—astronomers build the biggest dishes they can. The world's marquee radio telescopes are fearsome creations. The Robert C. Byrd Telescope in Green Bank is a full one hundred twenty feet taller than St. Paul's Cathedral. But telescopes like these won't do high-frequency work. Few telescopes are smooth, precise, and well situated enough to work at the highest frequencies of radio light.

A radio telescope's bowl-shaped reflecting surface is tiled with metal panels, each polished to exacting specifications. To accurately reflect one-millimeter light, for example, the panels must be free of bumps or scratches larger than one-twentieth of a millimeter.

With enough money, you can make enormous reflecting surfaces that are smoother than this. But there is never enough money.

High-frequency radio light creates other challenges. The sharper a telescope's resolution, the more accurately it must be aimed—"pointed"—at its target. Accuracy isn't a matter of being extra-careful turning the knobs and dials. The entire multimillion-dollar electromechanical apparatus that swivels and steers the hulking instrument must be engineered to higher tolerances. Such precision is expensive, so most telescopes don't have it. Big dishes also deform as they turn and tilt, because they're a little bit floppy. They also expand and shrink and warp depending on the temperature and time of day. You can install thousands of independently tweakable, computer-controlled actuators that continuously adjust each surface panel, keeping the telescope in focus, but, again: expensive. For all these reasons, radio telescopes that operate in the millimeter-submillimeter realm tend to be small—six or eight or ten meters in diameter.

Let's say money was no object. You could spend millions of dollars upgrading the surfaces and mirrors and receivers and electromechanical controls of every dish in an existing telescope array (say, the Very Long Baseline Array) and it still wouldn't work at the highest frequencies of radio light, because those dishes weren't in the right places. Earth's surface is a terrible place to collect cosmic microwaves, which are easily absorbed or scattered by water vapor in the atmosphere. At low enough frequencies, radio astronomers can almost observe in the rain. High-frequency radio telescopes must be built in the highest, driest places possible. There are fewer suitable spots on this planet than you might expect. A good site for a high-frequency radio telescope should be well into the zone where emergency oxygen tanks are required, but it should be flat enough to hold a structure the size of a Manhattan apartment building. If you have to ice-climb to the top, it won't work: a road, however treacherous, should go to the summit. The site also needs to be in a reasonably peaceful and friendly country where you can ship crates filled with equipment that is technically and politically sensitive; the U.S. government, for example, still controls the export of hydrogen-maser atomic clocks.

By 2004, those long-awaited high-frequency observatories in Hawaii, Chile, and Mexico were nearing completion, but they weren't there yet, and when they were finished they *still* wouldn't be ready for this job, because they weren't equipped to do Very Long Baseline Interferometry. The upgrade list varied from telescope to telescope, but in general, each site would need atomic clocks, new signal-processing equipment and data recorders that were still being designed, and invasive surgery to implant this new hardware. All of this would cost millions of dollars and take months if not years of work.

But, Shep told the crowd, a force of technological nature was about to intervene: Moore's law. Named for computing pioneer Gordon Moore, the law held that the density of integrated circuits would double every two years—in other words, computers would keep getting cheaper and more powerful. Shep was betting that the same relentless technological progress behind the iPod would transform high-frequency radio astronomy.

Moore's law wouldn't hurry along the construction of those new observatories, and it wouldn't turn a ten-meter dish into a one-hundred-meter dish. But it created a workaround. In astronomy, collecting as many photons as possible is the whole game. That's why big dishes are more sensitive than small ones—they can see fainter objects than small dishes can because they have more raw collecting power, more "steel." You can increase the sensitivity of a small dish by "integrating" for a long time—the equivalent of leaving the shutter open to take a long-exposure photo in dim light. But high-frequency radio light is so vulnerable to tiny fluctuations in Earth's atmosphere that integrating for a long time is like trying to take a long-exposure portrait through a pool of swirling water. Moore's law would help by ushering into existence affordable, powerful off-the-shelf microprocessors and hard drives that could replace creaky hand-built signal-processing equipment and slow, finicky reels of magnetic tape. Faster processors and higher-capacity recorders would increase the bandwidth at these smaller dishes, rapidly sampling and recording a wider range of light frequencies, making those instruments sensitive enough, Shep believed, to see Sagittarius A*.

• • •

Around the time those new digital signal processors and high-speed data recorders became available, Shep received a strong nudge. A few years back, a graduate student at the Harvard-Smithsonian Center for Astrophysics named Zhi-Qiang Shen had observed Sagittarius A* with the Very Long Baseline Array. The results, now being published in *Nature*, were good: Shen and his coauthors had seen farther into Sagittarius A*'s core than ever before. The *New York Times* announced those results with a headline declaring, "Astronomers Say They Are on the Verge of Seeing a Black Hole." Until now, Shep had been pacing himself through the first leg of a quiet, unsanctioned race in the dark. Now the floodlights were on. As the *Times* pointed out, and as Shep well knew, it would take new high-frequency telescopes to see the black hole's shadow. They weren't as close as the headline suggested. But it was time to move.

Without quite realizing what he was doing, Shep started building a team. Jonathan Weintroub was the first recruit. Jonathan worked for the Smithsonian Astrophysical Observatory, part of the Harvard-Smithsonian Center for Astrophysics, which occupies a compound of old academic halls and campus-expansion structures on Observatory Hill, west of Harvard Square. Like most of the radio astronomers, Jonathan's office was in an ivy-splashed brick building on Concord Avenue that the Center for Astrophysics rents from St. Peter's Parish next door. His job was to support the Submillimeter Array, the high-frequency radio observatory being built on the summit of Mauna Kea, in Hawaii, which was designed to achieve a thirty-fold increase in resolution over comparable telescopes of the day. Jonathan was a plush bear of a man in his early forties, with soft, curly hair and a casual, beachy manner. Everyone called him Jono. He spoke in a posh accent indistinguishable to most ears from that of an Englishman. He was friendly and eager to host yet blunt to a fault. When he and his family moved to the suburbs of Boston, a neighbor suggested they go get a beer sometime. Jonathan said, *No, I'm not really interested in socializing.*

Jonathan left South Africa in 1986 in an aboveboard effort to avoid the draft. He'd lived his whole life in the same house in Camps Bay, an affluent beachfront suburb of Cape Town, and he wasn't eager to go. But he'd just finished a master's degree in engineering, which meant he'd run out of legitimate excuses to defer compulsory military service. He tried to get assigned to an army research lab. Instead, he was ordered to join the infantry. Jonathan could see himself getting posted to a township, ordered to shoot tribespeople. So he flew to New York. If he periodically wrote letters that proved he was living out of the country, he could go home at a more peaceful time and face no legal consequences.

When he arrived in the United States, Jonathan got a job at a company in New Hampshire that made seafloor-mapping sonar for commercial ships. He was lonely out in the woods, so he started looking for a job in Boston. After a while, through an indirect connection, Jonathan ended up a doctoral student under Paul Horowitz, a revered Harvard electrical engineering professor. Because of Horowitz's side interest in building instruments to detect communications from extraterrestrials, Jonathan managed to insinuate himself into a career as an astronomer. When Shep came calling, Jonathan had just finished building the Submillimeter Array's correlator, which took the signals captured by each of the telescope's eight antennas and combined them. With that done, he was looking for a new project.

Shep and Jonathon had very different personalities and working styles. Shep was tightly coiled and intense. He tended to work in manic bursts, often through the night. He'd hop from task to task like a woodpecker flitting among trees, bashing his head against the object that filled his field of view. Jonathan was calm, deliberate, systematic, with the facilitating skills of a trained engineer.

In the complex across Concord Avenue from Jonathan's office, Shep found help from astrophysicists who were using novel theories and newly proficient supercomputers to build detailed models of Sagittarius A*. The don among them was Ramesh Narayan, a polymathic researcher who, among other things, studied black hole accretion—how black holes eat, and how their eating influences the universe around them. In 1997, he came up with a hypothesis to explain one of Sagittarius A*'s biggest mysteries—its dimness. The galactic center was filled with gas and dust and stars. A giant black hole in the middle of it should unavoidably eat a steady diet of that stuff, and that should make it shine more brightly than it did. No one expected Sagittarius A* to look like a quasar. But it was only about as bright as an average star; theory said that it should be at least ten thousand times brighter. Narayan argued that Sagittarius A*'s low luminosity was yet another reason it had to be a black hole: nearly all the energy people expected to see shining away in the form of light was going down the drain, disappearing behind the event horizon.

Narayan had attracted a flock of talented graduate students and postdocs, many of whom would eventually join Shep's mission—Charles Gammie, Dimitrios Psaltis, Feryal Ozel. Down the hall, at the newly created Institute of Theory and Computation, founded by Avi Loeb, a postdoctoral fellow named Avery Broderick was making models to explain evidence that Sagittarius A* wasn't just sitting there, semi-dormant: it was changing. New telescopes were picking up bright flashes of infrared light and x-rays, some lasting only seconds, that seemed to come from the edge of the black hole.

Avery and Avi ran computer simulations to explain these flares. They suspected that superheated matter swirling around the black hole was exploding into "hotspots," clumps that would orbit the black hole a few times before breaking up. X-rays crackled out of the hotspots as they circled the drain. The astonishing part was how fast the hotspots were orbiting. They might have been completing the twenty-four-million-mile lap around Sagittarius A* in as little as four minutes, suggesting that spacetime itself was swirling around the black hole—frame-dragging at work. To test these ideas, Avery and Avi programmed supercomputers to draw them pictures.

Simulating black-hole accretion is like modeling the weather in hell. When gravity becomes extreme, Einstein's equations metastasize. Because, as Einstein showed, mass is another form of energy, energy curves space just like mass does. And what else is energy? Gravity itself. That's right, gravity produces more gravity. Predicting the behavior of a system like this is a mathematical quagmire. But better computers and new methods were making things easier. When Avery and Avi ran their simulations, their computers drew scenes from inside a demented kaleidoscope. Glowing blobs swirled and lensed their way around the black hole. If astronomers ever managed to observe Sagittarius A* up close, Avery and Avi argued, they could see a flashing, churning maelstrom of matter and energy draining from the universe. Conceivably, if the telescopes got good enough, astronomers could take a quick burst of images and create a flip-book movie.

Avery and Avi explained all of this to Shep one day in Avi's plush-carpeted office at the Center for Astrophysics. Shep, in turn, explained that he intended to do this very thing. Avery was about twice Shep's size—a big, amiable man with fair, freckled skin and a long brown ponytail. As they sat and talked, Shep becoming intense with excitement, Avery thought to himself, What this guy is describing sounds like a lot of work, but people have been building worldwide telescope arrays for a long time. Why hasn't someone done this already?

The shortest path to Sagittarius A* began on the summit of Mauna Kea. There were three high-frequency radio telescopes up there, and

Shep wanted to use one of them to observe Sagittarius A* at the unreached wavelength of 1.3 millimeters, which was close enough to the magical one-millimeter mark to pierce the veil. The most desirable among them was the Submillimeter Array, but it still needed work before it could do VLBI. Shep could have delayed his first attempt by a year or two and used the SMA the first time out. But Shep didn't like to wait. Second choice was the James Clerk Maxwell Telescope, a big single dish next door to the Submillimeter Array. A distant third was the comparatively ancient Caltech Submillimeter Observatory.

Shep and his team got time on the Caltech Submillimeter Observatory. They had partners working at the Heinrich Hertz Telescope on Mount Graham, in Arizona. On the first clear night, when the Milky Way rose high enough above the Pacific horizon, they and their partners on Mount Graham initiated a prewritten sequence of telescope commands, and two silver dishes separated by five thousand miles quietly slewed in synch, following Sagittarius A*'s track across the night sky.

After several nights, they crated up their equipment and flew back to Boston. It seemed as if the observation had gone well, but it was impossible to know until the data was correlated. Practitioners of VLBI are blind nodes in a compound eye. At any moment, they can look at a monitor and confirm that their node is working. But resolving the source requires the entire compound eye. And if for any infinite number of reasons the light gathered at the different nodes can't be correlated into a composite picture, no one has seen anything.

Back at Haystack, they spent three months trying to correlate the recordings from Hawaii and Arizona. It would have been a lot easier if they had used three telescopes instead of two. With three telescopes, it's possible to isolate problems through process of elimination. If Hawaii and Arizona won't correlate, but Arizona and California will, then the problem is in Hawaii. With two telescopes, you guess.

They did a lot of guessing. Only after someone dismantled the receiver at the Hawaiian telescope did they learn that a broken component in a circuit board had ruined the entire observation.

• • •

Shep and Elissa now had two toddlers, a girl and a boy, and Elissa was finding that it was impossible for a family to have two frequent-flying, high-achieving academic parents without outsourcing some of the parenting. They didn't have nearby family to watch the kids, and Elissa wasn't an outsourcing parent. She'd seen plenty of shattered families here on the Harvard-MIT ambition farm, so she did what she had to do. Once she took an overnight trip and returned to have their son tell her, "Mommy, when you were gone, every minute of every day it was like those Dementors coming for Harry Potter when his mom was being killed." That was it for her. She'd keep her work travel to the minimum, whatever the impact on her career. She never ran the decision by Shep, who would have told her not to make any sacrifices for his sake. It was just what she needed to do to feel whole.

When she and Shep first got together, her mother had asked a question about her daughter's new boyfriend's profession: "Is there a need for that?" Elissa laughed but didn't answer. She wouldn't question the validity of a major branch of scientific inquiry. Yet on their first date, as Shep explained how every galaxy revolves around an enormous black hole, some of them as big as our solar system; as he swirled his hands to demonstrate the flow of an accretion disk and then pointed one finger at the ceiling and one at the floor to show how—one of the great mysteries is *how* this part happens—some black holes fire jets of matter out of their north and south poles at nearly the speed of light, redistributing matter across the universe, Elissa, in her mind, shrugged. *It's interesting, I guess?*

She understood the romance of the work. Before the kids were born, she went on lots of observations with Shep. One Arizona weekend when all the beds at the Heinrich Hertz Telescope were taken, they bunked at the Vatican's lodgings across the summit, quarters normally reserved for churchmen and scientists working at the Advanced Technology Telescope. The Vatican had established an observatory in Italy in 1891 to show that "the Church and her Pastors are not opposed to true and solid science, whether human or divine." A century later, against the wishes of the Apaches who consider the place holy ground, over the protests of environmentalists concerned for the endangered Mount Graham red squirrel, the Vatican joined the consortium of universities and science agencies behind the Heinrich Hertz Telescope and built their own instrument. In the mornings, Shep and Elissa talked about God and the cosmos with the Jesuit who cooked them breakfast and made them espresso. At night, the darkness was staggering. Flashlights and headlamps were forbidden because they would interfere with the telescopes. A mountain lion was prowling the summit that week. To get from building to building people had to cluster together and feel their way through the darkness, clapping and shouting to scare the big cat away. The chance of sudden death by apex predator heightened the sense of adventure.

So yes, she got it. And she found almost anything interesting once she started learning about it. But her exposure to astronomy was mediated through Shep, so she saw the whole enterprise in social terms, and she realized, earlier than Shep, that the field had a dark side. This awareness came to her in stages. She once went with Shep to a big astronomy meeting in San Antonio, Texas, where she noticed a couple of things. One: no women. Two: she was the only one asking questions. At one point, she tried to make conversation with a table full of young astronomers. So, what are you investigating? she'd ask.

I build numerical models of accretion onto neutron stars.

I study temperature fluctuations in the cosmic microwave background.

I am interested in the Epoch of Reionization and the large-scale distribution of structure in the universe.

Elissa would say, Okay, but what are you hoping to find out? Her impression was that at this stage in their careers, these people were so caught up in the mechanics of the work that they'd lost touch with what got them interested in the field to begin with. The clearest answer she got came when one young astronomer said, emphatically: *I want to know when time began!* The contrast with her own field could not have been starker. At the time, she was traveling from city to city, sitting in on intervention sessions in housing projects, watching prostitutes teach one another to cheek condoms onto unsuspecting Johns.

At observatories, the openness that made Elissa a good counselor and fieldworker elicited late-night confessions from overstressed, sleep-deprived men she barely knew. At every telescope she visited, she felt like the resident shrink. And she sometimes saw her would-be patients behaving badly. Just after dawn one morning at the Quabbin Observatory, to name an example, Shep and the other guys started arguing about whether to remove a reel of tape from the recorder. No one had slept in a day. The disagreement escalated into a shouting match and threatened to turn into something worse. Elissa genuinely thought someone was going to get punched. When deprived of sleep and subjected to pressure, a group of astronomers tends to combust.

Shep was comfortable working in this environment. He enjoyed calling himself an astronomer. In his mind, there was nothing remote or abstract or fussily intellectual about it: as job titles go, "astronomer" was all bullwhips and sidearms.

He got into this work because he was attracted to the mountaintops and the telescopes, the travel and the labor. Like anyone with training in physics, he held black holes in appropriate awe, but understanding them was not his primary motivation. Not at first. In time, though, something about black holes came to disturb and fascinate him. He thought that in a deep, existential way, black holes were creepy.

What really got him was this: the interior of a black hole is the only place in the universe from which it is impossible to return. This wasn't a case of "it would take a rocket containing the energy of all the stars in the universe" to escape. *No* amount of energy would ever get you out. In the equations that describe a particle falling into a black hole, when an object crosses the event horizon, the coordinate describing time takes on a "spacelike" aspect. In a sense, time becomes direction, and that direction points toward the center of the back hole. It is impossible to isolate the moment when you cross the event horizon, but once you do, your fixed path leads inexorably inward. The black hole becomes your future, and you can no more alter your fate than you can reverse the flow of time.

• • •

Shep had two ways of thinking about the failure of the 2006 experiment. The optimistic view: just get all the equipment working and they'd be fine. The darker one: what if the experiment was meant to fail the first time for a good reason, but instead it failed for a stupid reason?

Pretty much everyone outside Shep's burgeoning little team told him it was impossible to detect Sagittarius A* on a baseline as long as the one between Hawaii and Arizona, mainly because of an observation Thomas Krichbaum did in 1998, the data from which suggested that Shep's experiment was doomed. Usually the problem astronomers have is that their telescopes aren't sharp enough. Yet it's entirely possible to overresolve a source. A telescope must be matched to the thing it's meant to see. A radio interferometer that is too high in resolution for the job at hand—whose baselines are too long, and therefore, whose field of vision is smaller than the object being observed—is useless. It zooms in on such a tiny piece of the target that it doesn't see its edges; it misses the surrounding and delineating blackness of space. To a radio interferometer, which measures contrast, that is identical to seeing nothing. And if you assembled a VLBI array that was too big—whose telescopes were too far apart—you could look directly at a black hole and never know it. Is that what they had done last year?

Shep didn't buy it. He thought Krichbaum's measurements were wrong. And yet as they prepared to try again, those old results chewed at him. If their second attempt was ruined by some untraceable hardware glitch, they'd have a hard time getting a third try. Funders and telescope directors would be justified in telling them to give it up. They could be wrong. But Shep would never be able to prove it.

8

SPRING 2007

They'd need better telescopes this time, and they'd definitely need three of them. The Submillimeter Array still wasn't ready, but Shep managed to get time on the next-best thing: the James Clerk Maxwell Telescope, which had twice the collecting area of the observatory they used last year.

The Arizona station was, as always, on board. To complete the triangle, Shep was determined to recruit a new, long-awaited observatory in California's Inyo Mountains—the Combined Array for Research in Millimeter-Wave Astronomy, or CARMA. Technically, CARMA wasn't ready for normal observations, and it would take a lot of work to get one of its antennas instrumented for VLBI, but Shep tended to view details like these as minor inconveniences. He was developing what would become his signature move—agitating for time on telescopes before they're officially open.

Shep and Jonathan and Geoff Bower and others sent proposals for observing time to each of the three telescopes. They gathered much of the same equipment they had wrangled last year. They borrowed a maser and shipped it to CARMA. They packed racks of sleek black digital recorders and signal processors and dispatched them to the three telescopes. They were nearly ready to go. And then CARMA rejected their proposal.

This happened mere days before the observation was scheduled to begin. Shep called Geoff, who worked with CARMA, and said, *Geoff, you have to do something, we have to get on this telescope.* And Geoff said, reasonably, *What is there to do?* This is how the process goes. They were a bunch of junior scientists scraping together equipment to do an observation that no one thought would work. That's when Shep took a nervy step for someone without the oak-paneled clout of a Harvard professor: he called CARMA's director and asked him to intervene. Somehow, Shep talked him into it.

For the second consecutive spring, they spent a couple of weeks on Mauna Kea, installing borrowed equipment and testing that equipment and then waiting on the weather. On clear nights they'd stay up from well before dusk until after dawn, when they'd pack hard drives filled with billions of random numbers representing noise and cosmic signals into foam crates. They'd draw straws to decide who had to drive the crates down to Hilo and FedEx them back to Haystack for correlation. At the end of the run, they dismantled their equipment and shipped it back east. Then they all went home. They had no idea whether the experiment had worked.

• • •

A month later, Shep was sitting in his office at Haystack, logged in to the computer that was reducing the data from the observation. He was concentrating on an ASCII file filled with scan times, signal-to-noise ratios, baseline lengths, and other raw data, looking for evidence of a common detection between two of their three telescopes. He kept noticing a scan marked with the number seven. A signal-to-noise ratio of seven marks a threshold where the probability of a random noise fluctuation masquerading as a cosmic signal becomes very low. He got up from his desk and walked back to the correlator room to talk to Mike Titus.

The correlator room looked much like it did when Shep started his apprenticeship twelve years earlier. The old, tape-based machines were still there. Next to them stood the new, digital correlator, a black tower of blinking LEDs. Shep walked up to Mike's desk holding a printout of the detection. "That came through the other night," Mike said. "I'm suspicious."

Mike had been seeing some sevens that he knew were noise. This could be another one. But they looked more closely when Mike noticed that this scan looked a lot like recent detections of bright quasars, and he knew those were real. They ran an algorithm that Shep helped develop for his doctoral thesis; it averaged the data in a way that amplified real signals while suppressing spurious ones. The signal-to-noise ratio skyrocketed.

They had found a "fringe"—a common detection between two antennas. The term is a historical relic, a reference to the pattern of light and dark ridges, amplified crests and troughs, formed when two rays of light are coherently merged. When a VLBI array "gets fringes" on a source, it means everything worked: the antennas are in synch, seeing as one. Fringes are gold. Getting fringes means a toast is in order. Today, getting fringes meant that they had seen through the veil.

• • •

They spent the rest of the spring writing up the results. They hadn't gathered enough data to make an image, but they had seen *something*. On the other side of the veil was an "event-horizon-scale structure." Weirdly, that structure, whatever it was, was smaller than the expected apparent size of Sagittarius A*'s event horizon. What did that mean?

It was an exhausting time. Shep and Elissa were renovating a big Victorian fixer-upper they'd bought from their neighbor in Malden. They'd put the kids to bed, seal their bedrooms with plastic sheets, put on Tyvek suits, stay up late stripping lead paint, then go to work the next day.

At all hours Shep felt the mania and anxiety of having a career-making accomplishment in hand—if it didn't all fall apart. At work, he lost interest in his other projects. One day he walked into the office of his boss, Colin Lonsdale, director of Haystack Observatory, and said that he wanted to go full-time on Sagittarius A*. He wanted to quit working on LOFAR, a low-frequency radio array in a field in Holland that he and Heino Falcke were both involved in. It was time to focus. They'd seen the black hole at the center of the Milky Way! This was going to be big. But it would take all his concentration. Colin, a tall, bearded Englishman who speaks with soft authority in full, polished paragraphs, told him to do what he was going to do.

The closer they came to publishing their finding, the more anxious Shep became. At a meeting of the American Astronomical Society, he described their observation to a reporter, and afterward he was seized with fear he'd ruined everything: big peer-reviewed science journals don't accept results that have already been written up in the press. If that reporter ever wrote anything, no one noticed, so that fear never materialized. But when they finally submitted their results to *Nature*, Shep realized very late in the process that he'd misinterpreted some data—said, basically, that the small size of the "structure" they detected was because the black hole was spinning rapidly, when a more likely explanation was that the thing they detected was off to one side of the black hole's event horizon, swirling around in the accretion flow. When he realized his error his bowels turned to water. He was certain he had ruined his career. He emailed his editor saying he had to retract a portion of the paper. His editor asked him to revise a couple of sentences, and everything was fine.

The results ran in the September 4, 2008, issue of *Nature*. Both MIT and the Harvard-Smithsonian Center for Astrophysics invited him to give talks that fall. For a guy who nearly failed out of graduate school at the former institution and had yet to be accepted to the high-toned community of the latter, these were big invitations.

The Center for Astrophysics talk happened on November 20,

2008, in Phillips Auditorium. In this little old theater at the heart of the Center for Astrophysics compound, Shep explained his methods and results to the assembled professors and students. This was only the first step, he said. They'd seen only the faintest outline of whatever was hiding in Sagittarius A*'s core. With more telescopes, with better technology, they could sharpen that image. After the talk, as people cleared the room, a senior astronomer named Paul Ho walked up to Shep and asked him to elaborate on that vague gesture toward future plans. "So," he asked, "when are you going to image this thing?"

PART TWO

MONSTERS OUT THERE

9

SUBMILLIMETER ARRAY
MAUNA KEA, HAWAII
MARCH 19, 2012

"Whoa, there's something really wrong here," Shep said into his computer monitor. Just before sunset, he was pounding on a keyboard in the Submillimeter Array control room. Outside a wide bay of windows, eight glinting parabolic dishes stood in formation. Below the rust-red mountaintop, clouds spread to the horizon like a down comforter. The sky was edge-of-space purple, like the last thing experimental aircraft pilots see before they pass out. Patches of snow speckled the ground. The storm that deposited them several days ago had since traveled east, where it had been blocking the skies in California, thus delaying their three-station observing run.

The control room was a pressurized container of supplemental oxygen, keystrokes, and tense silence. "Well, it looks like we're actually recording something," Shep said. "Which is nice."

"The Mark 5Bs are recording," a visiting postdoc said. The Mark 5B recorders were the high-speed data recorders connected to the James Clerk Maxwell Telescope (JCMT) next door, which was contributing its fifteen-meter dish to tonight's effort. "The Mark 5Cs"—

the newest, highest-bandwidth recorders, and the ones hooked up to the Submillimeter Array—"are not."

Shep sprinted out of the room and ran downstairs, where the recorders were installed. A few minutes later, he darted back into the control room, panting in the thin mountain air. He was a trim forty-five years old. The hair that had not yet retreated still grew brown and long to form, late at night or in moments of acute stress, an impressive vertical pampas-grass tuft.

But it was still early. He sat back down at his computer, banged out a few keystrokes, and mumbled something reassuring to the postdocs and telescope operators. A little after 7 P.M., with a couple of minutes to go before they and their partners in Arizona and California began a twelve-hour sequence of scans on quasars (for calibration), Sagittarius A*, and the black hole at the center of the galaxy M87, the recorders seemed to be working.

Once the telescopes were running, Shep was calm. He sat down in an office chair and passed around a duffel bag filled with snacks he'd brought from Trader Joe's back in Boston. "You have to try the mission figs!" he said.

• • •

The vision that had come to occupy the minds of Shep and a growing mass of fellow travelers was to build, as he had taken to describing it, "the biggest telescope in the history of humanity." It would be a distributed Babel, constructed on as many as a dozen high perches around the world, from Europe in the east to Hawaii in the west, and, eventually, from Greenland in the north on down to the South Pole. It would have the highest resolution of any astronomical instrument ever assembled. It would meet the most crucial requirement of all: it would be able to resolve a doughnut on the moon, or, if you like, the shadow of the black hole at the center of the Milky Way.

Astronomers sometimes talk about an astronomical target's "view" of a telescope, and if Sagittarius A* were to develop sentience and look back, it would see a conveyor belt of silver dishes mounted on mountains, a sparsely mirrored disco ball spinning at the speed of night and day. First, the instruments in the Spanish Sierra Nevadas and the southern French Alps would roll by. A few hours later, a giant silver dish in Mexico and a cluster of antennas in the high Chilean desert would pass. Next, Arizona and California would appear. Hours later, Hawaii. Sagittarius A* would have eyes on the South Pole Telescope throughout the night. As Earth turned, these telescopes would view their target from many angles, accumulating perspective, filling in an abstract mathematical plane that a supercomputer would later turn into images. If all went well, one of those images would join the pantheon of iconic cosmic pictures.

When Paul Ho approached Shep back in 2008 after his talk in Phillips Auditorium, Shep was cagey. Of course he'd thought about "imaging," as they say, Sagittarius A*, and in talks and papers and proposals, he was comfortable making grand claims with strategic bravado. But one-on-one he'd get a little self-conscious, a little afraid of the jinx. It would take basically every millimeter-wave radio telescope on the planet to collect enough light, at high enough resolution, to make an actual image of the black hole shadow—if that was, in fact, what Sagittarius A* was hiding. There were no instructions for building an Earth-size telescope array from a bunch of observatories that were still under construction and controlled by international consortia with their own priorities and agendas. Even if he could put all those telescopes together, nature might be better at hiding her secrets than anyone expected.

But there was no way Shep could resist the pull. Together with his collaborators, Shep parlayed that first success into more telescope time. Each time they went out, they added some new capacity, reached some new goal, which they then wrote into next year's telescope-time applications and grant proposals. In 2008, they used the same Hawaii-Arizona-California triangle, and this time they got detections on all three baselines. A year later, they added the Submillimeter Array's eight brand-new, supersmooth six-meter dishes to their collection and spotted M87, aka Messier 87, aka Virgo A, a supergiant elliptical galaxy fifty-three million light-years from Earth and known to harbor a tremendous central black hole—one so large that, despite the distance, it should have a visible shadow.

The incremental successes compounded. In 2009, Shep led the submission of a paper to the Decadal Review Committee of the National Academies of Science arguing that it was "almost certain" that the "long-standing goal in astrophysics" of "direct imaging of black holes can be achieved within the next decade." The ten-page white paper concluded with assurances that "the path forward is clear." "Details of the technical efforts required to assemble this 'Event Horizon Telescope' will be described elsewhere, but no insurmountable challenges are foreseen." The committee included the Event Horizon Telescope in its list of national priorities for the coming decade.

More like-minded scientists joined the team every year. In 2007, a few months after the groundbreaking Mauna Kea run, a postdoc named Vincent Fish came to Haystack to do work on low-frequency arrays, and Shep conscripted him into his galactic-center pursuit. In 2009, Dimitrios Psaltis and Feryal Ozel got involved after meeting Shep during a visit to Harvard. Dimitrios and Feryal were both veterans of Ramesh Narayan's lab. They met there in the late 1990s, when Dimitrios, who grew up in Greece, was a postdoc and Feryal, who grew up in Turkey, was a grad student. They worked on a project together, and later they got married. Now they were both professors at the University of Arizona in Tucson, a good base of support for Shep and his endeavors. The university had just hired a stoic, overachieving young radio astronomer named Dan Marrone, who worked with the telescope on Mount Graham and with a single-purpose experimental dish at the South Pole. He was just as excited about staring down Sagittarius A* as anyone.

As a show of enthusiasm, in January 2012 the University of Arizona hosted a formal kickoff meeting for the Event Horizon Telescope (EHT) in Tucson. After that meeting, seventeen high-ranking professors and directors of observatories and institutes—"groups and facilities . . . who have taken part in 1.3 mm VLBI observations, who are making material contributions to the 1.3 mm effort, or who are enabling new EHT sites"—signed a letter of understanding that turned their years-long let's-put-on-a-show collaboration into an Organization with road maps and policies and the minimum level of necessary bureaucracy.

For the past four years, they had been carrying out annual observations with the same observatories: in Hawaii, the SMA and the JCMT; in Arizona, the Submillimeter Telescope, or SMT, (né Heinrich Hertz Telescope) on Mount Graham; and in California, the twenty-three dishes in California's Inyo Mountains belonging to CARMA. The plan for the next three years was to expand the array from three stations to eight, which would grow the collecting area of the array tenfold. Meanwhile, they would increase the bandwidth of the back-end electronics and data recorders from the current rate of one gigahertz to sixteen gigahertz. The jump in collecting area and bandwidth would enhance the sensitivity of the EHT forty times over. With that array, they believed they'd be able to get the first image of Sagittarius A*'s shadow. They'd do the first observations with the Earth-size telescope by 2015.

• • •

The weather on Mauna Kea was immaculate. Radio astronomers talk about the weather by invoking tau, a measure of the opacity of the atmosphere to starlight. The tau tonight was 0.028. Nights this clear came to Mauna Kea only ten or fifteen times a year. The observatory sat in a valley of volcanic cinders four hundred feet below the 13,800-foot summit of Mauna Kea, above half of the planet's atmosphere. Yet even at this altitude, home to little but Wekiu seed bugs and carefully acclimated humans never far from supplemental oxygen, the atmosphere is a constant nuisance. The clearest skies churn with microscopic turbulence.

In VLBI, good weather at one telescope means nothing if conditions are bad at the others, and tonight, the other sites were much worse off. The tau at CARMA was disturbingly high. The tau at SMT was excellent, but so far, airborne ice crystals had prevented the telescope operators from opening the dome. Passable weather at those sites might have to do, though. The snowstorms in California and Arizona had forced Shep and crew to spend the past few nights brooding five thousand feet below at Hale Pohaku, the dorm where astronomers stay while working on Mauna Kea. The observatories had given them three nights for this run. To increase the probability of getting good weather at all three sites, they could pick any three nights during an eight-night window. Tonight was their second-to-last chance this year.

Around midnight, Shep rose from his computer terminal, walked across the control room, picked up a landline phone—cellphones are forbidden because their signals interfere with the observatories—and called Arizona to find out when they'd be opening the dome. He hung up visibly cheered. "Yes!" he said. "SMTO's opening the dome and should be observing in about thirty minutes."

"Just in time to get two scans on M87," said a weary lanky guy in his mid-twenties named Rurik Primiani. Rurik was sitting at a monitor facing the control-room windows. He had the privileged-impoverished look of an Ivy League hipster: shaggy hair, thrift-store clothes, a vague air of expensive education. He was born in Caracas to an Italian-Venezuelan father and a Spanish mother. Both parents worked for Pan Am, and the family moved to Miami, one of the airline's big hubs, when Rurik was two. As an undergraduate at MIT, he studied engineering, but he also took a few astronomy courses, and those helped awaken the seeker in him. He decided to avoid industry after graduation, but he was skeptical about graduate school. He didn't want to work for free until his mid-thirties. So in 2008, he applied for a job doing engineering work for the Submillimeter Array. In the interview, Jonathan Weintroub handed him the group's big *Nature* paper on Sagittarius A*, which had just come out. The black hole drew Rurik to the job and kept him there.

Thirty minutes after talking to the crew in Arizona, Shep walked over to a phone and called them back, just to confirm that the dome was open and the station was online. He was quiet for several moments. "You're lying," he said. "No, you're lying."

"What did I break?" Jonathan said from across the room.

Five years after their breakthrough observation, Jonathan was just as determined to get a picture of Sagittarius A* as Shep, but his situation was different. Jonathan didn't work for the EHT: he worked for the Submillimeter Array, which spent only a few nights each year in service of the EHT. Hunting Sagittarius A* was a passion project that Jonathan tried to build into his day job as much as possible. But he couldn't devote every nanosecond of attention and fiber of being to it like Shep did. And obsession wasn't Jonathan's style, anyway. He fenced off portions of his life from the unending demands of work. He could make a Gantt chart to govern his giant long-term projects and set his pencil down each afternoon after completing that day's incremental work. And speaking of charts, he sometimes wondered why they, as a project, weren't better organized. They'd been observing with the same three telescopes for years now. When would they start growing?

Shep hung up the phone and explained that for unknown reasons, the Arizona station was not yet working. They were already on the twelfth scan of the night. Conditions in Arizona were fantastic—the tau there had dropped to 0.05, as good as it gets in the continental United States. After pacing around the control room for a few minutes, Shep called back for an update. "Now it's what?" he asked. " 'Going crazy'? Is that a technical term?"

A guilty snicker spread among the postdocs.

In two hours, Sagittarius A* would rise. The stakes tonight were higher than usual: NASA's Chandra satellite was joining in, watching Sagittarius A* for x-ray flares that, when combined with data from the EHT, could show how the black hole changes by the hour. So Shep exercised all the control he could from three thousand miles away. He asked the telescope operator to call the chief faculty member at the University of Arizona in the middle of the night and ask him to get there immediately. "Tell him, 'Shep threatened my life unless I called you.' "

Half an hour later, Shep received an email from Arizona and read it aloud: "There is 'no chance whatsoever' " that they would be back online tonight.

They faced a decision. It was still relatively early. They could cede the remainder of tonight to other astronomers. Or they could continue with the two-station array. They weighed the options.

Jonathan swiveled away from his laptop and said to Shep, "You've got the satellite coverage tonight from Chandra." Shep nodded. Satellite coverage is not something to squander. After a moment Shep said, "If Chandra detects a flare, we could do some very interesting science."

And after all, they were already on the mountain. The station in California was observing. They were running out of nights. And so the observation continued, with the first scan of Sagittarius A* scheduled for 2:05 A.M. Decision made, Shep collapsed into an aluminum folding chair.

By 2:30 A.M., two-thirds of this early demonstration version of the Event Horizon Telescope was recording transmissions from Sagittarius A*, which hung low over the horizon.

Shep leaned back in an office chair and closed his eyes. Jonathan lay down on the floor and fell asleep. Everyone else kept monitoring their computers. Two and a half hours passed in which nothing happened, which is the way these things are supposed to go.

By 5 A.M., everyone was awake, and Rurik, still sitting behind his control monitors, was getting restless. "Think we have enough data now?" he asked Shep.

"The question is whether we're getting any data," Shep said. "Who knows what CARMA's doing. Pretty sure we know what SMTO's doing."

A little after six, Shep roused the postdocs and prepared to power down the machines. They placed eight eight-terabyte hard drives of almost certainly marginal data in a foam crate and packed it into the back of a truck, then drove down the summit road to the base camp, squinting in the morning light. Over breakfast in the cafeteria at Hale Pohaku, Shep gave a pep talk, which began: "Okay, guys, that was a disaster."

10

By big-science standards, the Event Horizon Telescope was a bargain. All told, Shep needed to raise maybe $20 million, and the return on investment could be significant. A picture of a black hole and the incidental observations made in its pursuit could help answer a long list of hard questions. The experiment could reveal whether general relativity holds in extreme environments where it has never been tested. It could determine whether Roy Kerr's metric for rotating black holes describes real, physical objects. It could show whether event horizons really exist. It could test the no-hair theorem, the idea that a black hole can be described by just its mass, angular momentum, and electric charge, and the cosmic censorship conjecture, which holds that singularities—the knots of spacetime thought to reside at the centers of black holes—can never be "naked": they're always hidden by an event horizon. If the shadow appears, cosmic censorship holds. If not, then Sagittarius A* might be a naked singularity—incomprehensibility made manifest and exposed to the rest of the universe. And, of course, the experiment might just produce an iconic astronomical image. There's no downplaying the significance of that. As Avery Broderick took to saying, the first picture of a black hole could be just as important as *Pale Blue Dot*, the photo of Earth that the space probe Voyager took from the rings of Saturn, in which our planet is an insignificant speck in a vast vacuum. Their picture, Avery thought, would have a different message: it would say, there are monsters out there.

But $20 million is still a lot of money, especially when you don't have it. Shep had always relied on funds from the National Science Foundation, or NSF, but government money was getting tighter. The world hadn't recovered from the recession that began with the financial crisis of 2008. The loudest people in America were still angry that the government had spent hundreds of billions of dollars bailing out banks. And it was an election year. These conditions turned an already competitive funding environment into the Hunger Games. To get NSF money, a project not only had to be impressive and important and generally meritorious—it, and the people behind it, needed to present a near-flawless track record. Hence the pressure that night on Mauna Kea.

Part of the funding game was publicity, and Shep was a good pitchman. After the *Nature* paper was published in 2008, Shep began to receive a stream of media requests that were high profile enough to be intimidating and intermittent enough to prevent him from getting comfortable in front of the camera. Once, he found himself under the lights of a BBC documentary film crew, gesturing and telling an interviewer, with coached passion, that if they take a picture of Sagittarius A*'s shadow, "that will be the *money shot.*"

Cut!

Er, Shep, could you try that again with a phrase that is less . . . pornographic?

Shep agonized over grant proposals until moments before deadline. Everyone agreed that he was good at writing them. He'd learned how to gloss over some of the Event Horizon Telescope proposal's weaknesses. It hurt that the project was designed mainly to observe just two objects—Sagittarius A* and the black hole at the center of M87. The EHT was a little like a Zeiss macro lens you'd buy to take pictures of flowers on a trip to Costa Rica: tempting, but worth the money?

Sharp-eyed proposal reviewers would notice logistical challenges. These guys were having trouble finding time to observe when the weather was good at three sites. How would they fare when their work required clear skies at eight or nine observatories spread across four continents? And what happened if one of those observatories closed? Astronomy was increasingly a zero-sum game: to build a new telescope, you often had to close an old one. The Event Horizon Telescope wouldn't work without the simultaneous participation of at least seven observatories, some decades old and some still under construction. Funding could shift even among telescopes within the EHT. Some people worried that the Atacama Large Millimeter Array would put the Submillimeter Array out of business. That would be a natural, resource-preserving move—unless you needed both telescopes to see what you wanted to see.

Shep had a few strategies for dealing with this grim litany. The first was to embrace the parlay method that had gotten them this far—to turn light into data into findings into grant money. The second was to hustle. The people who built and ran telescopes tended to think in cosmically extended time frames. What difference is a year or two in cosmic time? For Shep, delays were death.

Strategy aside, Shep was counting on something else. Every time the weather was good, every time the telescopes failed to malfunction, the EHT crew had to give partial credit to luck. This was certainly true in early March 2012, when Shep opened an email from someone who worked for the Gordon and Betty Moore Foundation.

• • •

Dusan Pejakovic of the Moore Foundation learned about the Event Horizon Telescope from the journal *Science*. The article was a news report from the kickoff meeting in Tucson, and it made the project sound like the kind of rare bird grantmakers like to collect. It was sexy, international, technology driven, affordable, seemingly impossible, yet apparently tractable, and, also, in need of help. Shep admitted to *Science* that they were still short on cash.

Dusan's email was well timed. The EHT was running on the fumes of a three-year grant from 2009, and there was no guarantee that their next NSF proposal would be approved. Outside money would power the project through the next couple of years. Shep and

Dusan talked on the phone, and Shep started sketching out a formal proposal. Then he absorbed a blow from the administrators at MIT.

Usually when a scientist gets grant money, the university takes a cut, often around 30 percent, to cover overhead—electric bills, lawn mowing, janitors. The Moore Foundation offered to pay 15 percent of overhead, no more. When MIT refused the terms, it initiated Shep's slow-motion departure from Haystack Observatory.

If Shep were faculty at MIT—if he had an oak-paneled office—he would have had more leverage. But Shep was a staff scientist. He was of a lower rank. And so he found himself sitting in his car in a Whole Foods parking lot along the Mystic Valley Parkway, on the phone with MIT administrators, stomach imploding, incredulous at hearing that one of the world's great universities didn't want to deal with the inconvenience of receiving a $1.8 million grant. He kept trying to make it work at MIT, but when those efforts failed, he called up the Harvard-Smithsonian Center for Astrophysics, which said, *Welcome aboard!*

The terms of the deal were that Shep would work half-time for the Smithsonian Astrophysical Observatory and half-time at Haystack. In return, Shep would get a salary, an office at the Center for Astrophysics (CFA), all overhead costs covered, and discretionary start-up money. The expectation was that after a few years, Shep would go full-time at the Smithsonian Astrophysical Observatory. And so in December 2012, Shep moved approximately no belongings into an empty yellow-beige office on Concord Avenue one floor down from Jonathan Weintroub.

The postdocs the Moore Foundation paid for arrived the following September. Their names were Laura Vertatschitsch and Michael Johnson.

Laura introduced herself to the CfA community with an acoustic guitar ballad called "Instrumentation Improvements for the Event Horizon Telescope, or, A Love Song for Black Holes." Postdocs had been invited to present their research with a standard science talk or a three-minute "haiku talk." Laura interpreted the haiku format liberally. Earlier on the day of her performance, in the first-floor office on Concord Avenue the new postdocs shared with Rurik, Jonathan helped her prepare. As she strummed her guitar and studied her PowerPoint slides, Jonathan looked at her with a mixture of awe and concern. "You know," he told her, "you're really very brave."

It was more like Laura was incapable of fear. She was boisterous and relentlessly optimistic. She had huge amber eyes and probably could have taken any of her colleagues in an arm-wrestling match. She grew up in Seattle, where her father worked as an engineer for Boeing. She was the oldest of nine kids—seven sisters and a brother. She was a lifelong athlete who started taking karate when she was seven and went on to compete in world-cup championships in Europe and South America. A TV channel for kids called Wham! once filmed her at a national karate tournament in her braces and slicked-back sports hair for a three-minute motivational story. *Stay in school, kids!*

She came to the EHT with a doctorate in electrical engineering and an expertise in advanced radar systems—specifically, a component of those systems called the Field-Programmable Gate Array, or FPGA. Used to be that if a scientist needed some bespoke computer chip, she'd design it and then hire a company like Texas Instruments to cast it into silicon. This process is just as slow and expensive as it sounds. For a lot of applications, FPGAs, which could be reprogrammed at will, made that process unnecessary. High-performance FPGAs were one of the advances that made it possible to build custom signal-processing equipment for tens of thousands of dollars rather than millions. Laura was here to help build the next generation of that equipment. She saw the ad for the postdoc job as graduation approached. She was intrigued by this crazy-sounding science project, and she thought, *I know how to do exactly what they need.*

She was unaccustomed to the forced priestliness of the astronomical community, though, so when she took the stage in Phillips Auditorium and looked out on a roomful of Ph.D.s wearing their science faces, she started, uncharacteristically, to sweat. Then she went through with it. Later, someone told her that her talk was the only one anyone remembered.

Michael Johnson, the other new postdoc, was a theoretical astrophysicist from the University of California, Santa Barbara. He was a mathematician at heart, but he spoke in complete paragraphs and didn't hesitate to use words like "beautiful" and "extraordinary" and "wonderful" to describe scientific concepts and accidents of nature. He had sandy brown hair, soft features, and the mildly bashful manner of someone a little embarrassed at his own earnestness.

In his studies, Michael was drawn to the extreme, and one day in grad school when he saw Dimitrios Psaltis give a talk on the Event Horizon Telescope, he felt the old worldline bend. Michael found most talks to be marginally interesting at best, incomprehensible at worst, but Dimitrios's lecture made such an impression that Michael reached out to Shep, and the next time Michael was in Cambridge, they had lunch.

Shep hired Michael because he needed someone with theoretical chops to analyze the data piling up from their annual observations. Michael came ridiculously well recommended. His mentors compared him to a young Ramesh Narayan—someone smart and well trained enough to handle just about any problem in theoretical astrophysics. Behind his back, Shep called Michael "Magic" Johnson. As soon as Michael arrived, Shep put him to work on data that had been sitting around since the spring 2013 observation, when the EHT made a priority of collecting polarized light. They wanted to use that light to map the magnetic fields in the galactic center, which would allow them to explore the long-standing mystery of how black holes eat.

It's not easy to fall into a black hole. Bodies in a stable orbit around a gravitating object, even one with a reputation as an insatiable annihilating force, tend to remain in orbit unless disturbed. The planets in our solar system have been sailing along in their current orbits for more than three billion years, and they'll probably stay in roughly the same orbits for another five billion years, until the dying, swelling sun absorbs Mercury, Venus, Earth, and Mars.

The accretion disks around black holes are made of plasma heated to billions of degrees—a thin, runny soup of widely spaced electrons and ions. These particles, like planets, should, if left uninterrupted, keep sailing along, circling the black hole. Sometimes they bump into one another, but that shouldn't be enough to knock them out of orbit and send them spiraling downward. Some other source of friction must be involved. The Soviet theorists Nikolai Shakura and Rashid Sunyaev proposed in 1973 that "turbulence"—big, stormy, disruptive collisions within the accretion flow—was the cause. But what creates the turbulence?

In 1991, Steven Balbus and John Hawley found a likely candidate. Magnetic fields are ubiquitous throughout the universe. Stars, galaxies, and planets with molten cores all generate their own magnetic fields. The dust and gas that makes up the interstellar medium has become magnetized over the eons by exposure to stellar winds, supernovas, and other forms of electromagnetic violence. It follows that the plasma swirling around black holes is magnetized, too. Theorists have shown that magnetic field lines should run through black-hole accretion disks like invisible thread. These field lines bind particles together like springs. As the gas in the disk rotates around the black hole, the magnetic field lines move, too. They become twisted and tangled, and sometimes they "reconnect" in a violent event much like a solar prominence. In this way, magnetic fields should create a disruptive churn, creating the viscosity that sends matter down the drain.

These magnetic instabilities could also explain cosmic jets. It's possible that after magnetized gas falls into the black hole, crossing the event horizon, its magnetic field lines remain. As the black hole spins, it winds these magnetic field lines up like an electric motor, extracting energy from the black hole and launching ruler-straight jets of energy hundreds of thousands of light-years into intergalactic space.

The magnetic instability hypothesis was largely untested, because there had never been a good way to map the field lines near the edge of a black hole. That's where the polarized light the EHT collected in 2013 came in. Light has two components: vibrations in the electric field, and vibrations in the magnetic field. In unpolarized light, the orientation of the electric field vibration is random. In polarized light, it is confined to one direction. The microwaves emitted by Sagittarius A* are inherently polarized, because they are emitted by electrons spiraling around magnetic field lines. The direction of the polarization in the light should tell you the direction of the magnetic fields. Were they pointed in random directions, or were they linked up in orderly patterns like the best theories of black-hole accretion predicted? Michael Johnson's first job at the EHT was to figure this out.

The office that Rurik, Michael, and Laura shared was big, bright, and spare. Michael and Laura sat facing opposite walls. Rurik's desk faced the center of the room, like that of a supervisor.

Tripling the number of people in this room was a morale boost for everyone. The project needed to grow to succeed. They needed more money, more telescopes, and more people. But as they were about to learn, resources come at a cost.

11

SANTA FE, NEW MEXICO
WEEK OF SEPTEMBER 29, 2013

Maybe a few hundred people worldwide studied the galactic center for a living. Most of them were in Santa Fe the last week of September 2013 for "The Galactic Center: Feeding and Feedback in a Normal Galactic Nucleus."

Conferences like this are supposed to be an opportunity for scientists to share findings and engage in high-minded debates, and some of that happens. But these meetings are also theaters of political operation. The astronomers who descended each morning on the big adobe hotel on Santa Fe's old plaza, some straight from their morning hikes in dusty fleece and Gore-Tex, were the peers who would referee Shep's scientific papers, score his grant proposals, and otherwise exude or withhold confidence that the Event Horizon Telescope was a going concern. Shep knew he had to project confidence in his ambitious timeline—first observations with the full, planet-spanning telescope array by 2015—even when, say, Jim Moran, who knew more about VLBI than anyone on Earth, concluded his Wednesday morning talk in Santa Fe by saying he was optimistic that the EHT would work, "though it might not be in my lifetime." A few minutes later, Shep wondered aloud, "Does he have terminal cancer?" And then Shep had to strike just the right tone of forced openness when Heino Falcke told him that he and a couple of European colleagues had submitted a mammoth funding proposal to the European Commission asking for fifteen million euros to do exactly what the Event Horizon Telescope was planning to do.

• • •

When his big 2000 paper on the shadow of Sagittarius A* was published, Heino was back in Bonn, working as a research scientist at the Max Planck Institute for Radio Astronomy. The institute was forty-five minutes from his hometown of Frechen, the Cologne suburb where his family had lived since the late seventeenth century. They had been active in the Protestant church in the central village for just as long. Heino was a lay minister there. He saw no conflict between the spheres of Luther and Einstein. In fact, he had a biblical analogy for black holes: the plight of Lazarus. *Before Jesus brought Lazarus back from the dead,* Heino said, *he was trapped on the other side of death, looking out at the world of the living. Inside a black hole, one would experience something similar. Escaping would be a miracle on par with resurrection.*

After that 2004 meeting at Green Bank, when Heino and Geoff Bower and Shep pitched the attendees of Sagittarius A*'s birthday party on taking the thing's picture, the three started holding semimonthly conference calls—"telecons," as astronomers say—to feel their way toward this elusive goal. The minutes from their discussions convey the good-natured cluelessness of kids trying to start a rock band. "Perhaps we should keep a list of action items and take turns with writing minutes," concludes the first installment. They made some progress. Shep and Geoff started working on signal-processing equipment with a group at Berkeley that built open-source hardware and software to search for extraterrestrial life. They scraped together money to refurbish antique masers.

But after a while, for no single reason, the proto-collaboration drifted apart. Heino argued that they should create a formal science collaboration modeled on big organizations like the Large Hadron

Collider (LHC). The LHC had spent billions searching for the Higgs boson, and they weren't anywhere near finding it at the time. Surely, he thought, they could get a few million dollars to image an event horizon. But he sensed that Shep was more interested in doing his own technical development work than in founding a big consortium. He'd come to think that the telescopes and technology they needed were still years from perfection. Heino decided that while Shep and the people at Haystack were doing that perfecting, he'd focus on building up his reputation as a scientist.

Heino pursued this goal like an investor building a diversified portfolio. He kept up his studies of black holes and quasars, but he also became project scientist for LOFAR, an array of low-frequency radio antennas in the Dutch countryside designed to peer back into the cosmic dark ages, before the stars lit up. He got involved with the Pierre Auger cosmic-ray experiment. He did some work on the search for extraterrestrial intelligence and advocated putting radio telescopes on the moon. In 2007, the same year Shep and company made their first breakthrough observation of Sagittarius A*, Heino took a job as a professor of astrophysics at Radboud University in Nijmegen, the Netherlands. From that base, he became a Dutch academic star, writing newspaper columns and giving interviews and winning awards. In 2011, for his work on the black-hole shadow and LOFAR, he landed the Spinoza Prize, Holland's highest scientific honor. It came with a roughly three-and-a-half-million-dollar cash prize.

He dangled this money in front of Shep at the foundational Event Horizon Telescope meeting in Tucson in 2012, after which the seventeen representatives of "groups and facilities . . . who have taken part in 1.3 mm VLBI observations, who are making material contributions to the 1.3 mm effort, or who are enabling new EHT sites" signed the letter of understanding on which the EHT was built. He said to Shep, *I can use this, tell me what we need to do.* He didn't understand why Shep didn't take him up on the offer—why it didn't get his name on that letter of understanding.

When the call for proposals to the European Research Council

Synergy Grant program went out in the summer of 2011, it was estimated that eight hundred groups would compete for about a dozen awards, putting the odds of success at about one and a half percent. But hey, why not? Heino was approaching the end of a few projects and needed to figure out what to do next. So was Michael Kramer, a pulsar expert at the Max Planck Institute for Radio Astronomy. These "synergy" grants were intended to create unusual collaborations. With Luciano Rezzolla, a theorist who specialized in mathematically modeling gravitational waves, Heino and Michael Kramer came up with a plan to observe black holes, find pulsars orbiting those black holes, and feed data about these systems into computer simulations that would test general relativity and competing theories of gravity. Heino's portion was devoted to a project he called BlackHoleCam. It would coordinate a worldwide network of radio telescopes to capture an image of the shadow of Sagittarius A*. They submitted their proposal on January 10, 2013, the day of the deadline.

Heino had to admit, he didn't really coordinate this proposal with Shep. He figured, we'll do it on our own, and it's crazy anyway. It's not gonna fly. But then the proposal kept advancing through the process, clearing round after round, and by the time of the meeting in Santa Fe, Heino's odds had improved significantly.

• • •

Shep's talk was scheduled for Friday afternoon, the last day of the weeklong conference, when plenty of attendees would already be shuttling toward the Albuquerque airport. It was some consolation that throughout the week, the words "Event Horizon Telescope" regularly came up in talks by people Shep hardly knew. The project was entering the community consciousness.

Shep walked to the front of the conference room, queued up his slides, and went through the usual drill—the theoretical background, the progress they'd made since the early 2000s, the plans for the next two to three years. He explained why it was important to stick to this schedule—they had to do this observation during the potentially slim window in which all the telescopes they needed were still in business. "ALMA puts a little bit of pressure on existing VLBI sites," he said. "We need to kick this off asap so we don't lose any critical sites." And then he played the G2 card.

To be fair, everyone at this conference had played the G2 card. Astronomers had discovered a cloud of gas roughly three times the mass of Earth barreling toward Sagittarius A*. On its current trajectory, it would make its closest pass within months. Everyone expected Sagittarius A* to rip the cloud apart. They could see a black hole eating in real time. Naturally, every astronomer with an interest in the galactic center was using the G2 cloud as an excuse for telescope time. The G2 cloud, Shep said, was a "once-an-eon" opportunity. Problem was, it was starting to look like G2 might be a dud. It seemed as if every telescope in the world was watching the cloud travel along its trajectory of doom, but so far, no one had seen any fireworks.

When it was time for questions, the audience of Shep's peers lasered in on the Event Horizon Telescope's vulnerabilities.

"How many of the U.S. sites do you need?" someone asked. In other words, if one of the telescopes you're counting on shuts down, are you screwed?

"It's a Sophie's choice kind of thing," Shep said. "We're robust against losing one or two. But I think we'll have enough sites."

Then came the really hard question: how much money do you still need?

"How much you got?" Shep replied.

• • •

A few weeks after the conference in Santa Fe, Heino, Michael Kramer, and Luciano Rezzolla walked into a twenty-four-story glass tower in Brussels and took an elevator to the top. They'd made the final round of the European Research Council grant competition. At this point they had a 40 percent chance of success. All they had to do was pass the final interview.

They had choreographed their performance to express synergy

through shared speaking roles and body language. The day of, they paced in the waiting room until it was their turn, and as they were led down the aisle into the U-shaped auditorium, Heino felt like a gladiator stepping into the arena. They flubbed a question about the budget, but that only cost them a million euros. The other fourteen million were theirs.

Heino insisted that he never meant to compete with Shep. It said as much in the press release, below the photo of Heino, Luciano, and Michael posing like a jazz trio on the roof of the Max Planck Institute for Radio Astronomy. It was the golden hour. Cirrus clouds streaked the clear blue sky. They wore blazers and button-down shirts, no ties. "Falcke first proposed this experiment 15 years ago and now an international effort is forming to build a global 'Event Horizon Telescope' to realize it," it said. "The BlackHoleCam team will collaborate with the Event Horizon Telescope project, led by Shep Doeleman," it said.

• • •

As a scientist in the field and a human being with free will, Heino was entitled to start his own, competing campaign to image Sagittarius A*. But BlackHoleCam wasn't exactly a competitor. Heino didn't have agreements with the telescopes he'd need. Shep did. The success of BlackHoleCam was contingent upon merging with the Event Horizon Telescope. From Shep's perspective, it was as if Heino had invited himself to join an Everest expedition by offering to help with the bills, and yet Heino, despite his history of studying Sagittarius A*, had never climbed a mountain.

Heino saw things differently. He knew the planet wasn't big enough for two Earth-size virtual telescopes. He knew when he wrote the European Research Council grant proposal that if it was successful, he and Shep would have to work together. But he held a collectivist view of big scientific experiments. Shep needed the money: why would he not want to take it from Heino? Heino was a serious scientist who had been writing about the prospects for imaging Sagittarius A* for his entire career. Why would he not get involved? Heino was generally on message about his motive for pursuing the competing ERC grant, but he'd also admit to feeling excluded, even disinvited, from the official circle of signatory EHT members. Heino's grant-based chess move might also have been a shrewd response to historical precedent. If—a big if—the Event Horizon Telescope were to win an award on the level of the Nobel Prize, only two or three of the hundreds of people who contributed to the project would get their names on it. Few scientists did more than John Wheeler to bring the concept of black holes into the world, but who won the first Nobel Prize related to black holes? Riccardo Giacconi, the scientist behind Uhuru, the x-ray telescope that marked Cygnus X-1 as the first likely real-life black hole. If the EHT went on to glory, Heino wouldn't get much credit for simply writing a paper back in 2000 suggesting that such an experiment could be done. He'd have to be actively involved in the experiment himself. He'd have to find a way in.

In December, Heino flew to Boston to talk to Shep and others about combining their efforts. This first meeting was perhaps the most cordial one they'd ever have. Shep hadn't yet had time to stew.

Shep was leading the Event Horizon Telescope because he decided after the first big observation in 2007 to make the pursuit of Sagittarius A* the sole focus of his career. As a staff scientist at a federally funded observatory, he didn't have a clear path to the security and prestige of a tenured professorship. That was the bad part. The good part was that he was free to pursue one seemingly insane goal with a monomania unavailable to tenure-track professors. He didn't need to build a portfolio. He could bet his career on a single experiment. He'd just have to protect his ownership stake in that experiment with the appropriate level of zeal.

But he also knew he had no choice but to absorb Heino's effort into his own. They could use each other. It wasn't so much about the money. It was true that Shep needed more funding, but Heino didn't have carte blanche to spend fourteen million euros as he liked—only a third of that went to BlackHoleCam, and much of that portion was earmarked for specific purposes, like hiring postdocs in Europe. Besides, Shep was almost finished with a proposal to a National Science Foundation program called the Mid-Scale Innovations Program (MSIP) for nearly $7 million, which would fund the EHT through the first big full-array observation. No, the real reason Shep had no choice but to let Heino in was that the Atacama Large Millimeter Array, or ALMA, the king of this realm, would, if only implicitly, insist. ALMA was run by a three-way partnership between science agencies in Europe, North America, and Japan. With the European Research Council behind him, Heino had the support of one-third of this trinity—enough to stop Shep from getting on the telescope that was the centerpiece of his entire plan.

12

HAYSTACK OBSERVATORY
FEBRUARY 6, 2014

ALMA was the world's most powerful telescope of its kind and, arguably, the most complicated astronomical instrument ever built. It consisted of sixty-six dishes planted on a virtually lifeless 16,600-foot plateau in Chile's Atacama Desert. It was the merger of three different dreams dating to the 1980s, when astronomers in North America, Europe, and Japan separately drew up plans for huge arrays of millimeter-wave radio telescopes that they would use to map the galaxy, watch planets congeal out of rock and ice, and hunt for the very first stars. All three groups wanted to build their telescopes in the Atacama, which outside Antarctica is the driest place in the world. Building three nearly identical observatories in the same place was ridiculous, so eventually, these efforts merged.

Starting in 2003, scientists and engineers and local workers spent a decade leveling desert ground, laying fiber-optic cable, grading extra-wide roads, and pouring 193 six-foot-deep triangular concrete pads for 66 movable parabolic dishes. The dishes would be placed in different configurations for different purposes: tightly clustered for sensitive studies of cold, dim things, or extended as wide as eleven miles for high resolution on small, faraway things. As with all big telescope projects, construction took years longer than expected. Finally, in March 2013, the Chileans and their international partners summoned dignitaries and journalists to the high Martian desert for inauguration. During the ceremony, the dishes slewed in time to music. Pierre Cox, director of the observatory, announced that with this instrument, astronomers had already discovered new and unexpected phenomena, including galaxies that were growing baby stars a billion years earlier than expected.

If Shep could bring ALMA into the EHT, it would unite the rest of the telescopes around the world and boost the sensitivity of the whole array tenfold. Without it, their chances of getting a picture of Sagittarius A* were slim. That's why he had been agitating for time on the telescope for years, and why he and a team from Haystack were in year three of a five-year project to turn ALMA into a phased array—a bunch of antennas that work together as one giant dish, much like the EHT itself. The way ALMA was designed, each of its sixty-six dishes captured its own images and recorded its own data stream. If you tried to do VLBI with ALMA in its current state, when it came time to correlate the data from Hawaii, Arizona, Chile, and all the rest of the telescopes around the world, ALMA's data would show up as sixty-six caches of hard drives from sixty-six individual telescopes. It would be a computational nightmare. The EHT scientists needed ALMA to pretend it was a single giant dish: to continuously sum the data recorded by all sixty-six antennas into a single master output. To make that happen, they had to perform surgery on ALMA's correlator, the most powerful single-purpose supercomputer in the world.

The process for modifying ALMA is only slightly less onerous than the procedure for carving a new face into Mount Rushmore. As a global collaboration of publicly funded science agencies, and as a very new, very powerful, very in-demand instrument, ALMA was governed by laws designed to ensure that no one gets special treatment. After your upgrade has been vetted and formally accepted, it must be advertised as a capability available to astronomers everywhere, and to use it, you get in line and submit a proposal like everyone else.

Shep had always felt he had a gentleman's agreement with the relevant powers. His group was giving ALMA a multimillion-dollar upgrade, and his project was urgent, because the EHT had to get running before any of the telescopes it needed closed—a possibility that ALMA's existence made more likely. For these reasons, he'd get to enter through the VIP entrance: he'd get to apply for director's discretionary time or cut into Cycle Two, the official block of telescope time scheduled for spring 2015. Only one man controlled access to these precious resources: Pierre Cox. But lately Shep had been receiving mixed signals from Pierre. Unexplained silences, noncommittal replies.

Shep's job this morning was to secure his special access. He drove to Haystack in his usual state of controlled panic, eighty miles per hour on Interstate 93, sheaves of snow calving from the unswept roof of his dilapidated Honda CRV into the paths of other drivers. When he got to Haystack, he summoned Mike Hecht and Geoff Crew into his office to prepare for the phone call.

Mike Hecht, an assistant director at Haystack, managed the ALMA Phasing Project. He was an assuring man in his fifties who projected the easy confidence of the extremely well organized. Geoff Crew, the scientist on the ALMA Phasing Project who made most of the working trips to Chile, wore his long brown hair in a ponytail and had the easygoing, hey-man bearing of an old hippie. In their different ways, Mike and Geoff both knew how to deal with Shep when he was in one of these frantic states. The key was patience.

Shep sat down behind his desk, and Mike and Geoff pulled up rolling office chairs. Shep intended to extract as much intelligence as he could in the five minutes before his call with Pierre. Geoff spent the most time at ALMA, so he got the most urgent grilling. "Who did you have a good working relationship with there?" Shep barked at Geoff. "Give me their names."

"There's Sue—"

"Sue who, I don't know Sue. Who else?"

They batted names and acronyms back and forth like Olympic table-tennis players: *CSV APP JAO version 10.4 versus 10.6 Cycle 3 Cycle 4 TAC*. At one point, Shep mentioned that they were using "last year's acronyms." And then Mike and Geoff were dismissed. Shep closed the door and waited for the phone to ring.

During the call, Pierre, a Frenchman living in Chile, was polite and diplomatic and impossible to read. Shep hung up the phone, unnerved. Shep and Pierre went back awhile, to the days when Shep was starting out with high-frequency VLBI and Pierre was the director of the IRAM Thirty Meter Telescope. They had a good rapport based on the understanding that Shep would try to get everything he possibly could from Pierre, and that Pierre would smile and parry and protect his own interests. Their negotiations were friendly sparring matches. But now, Shep wondered why Pierre seemed to be walking away from what he'd always thought was a guarantee of director's discretionary time. Why was the deal unraveling? Shep had a schedule to stick to, which put them in position to observe with the full globe-spanning telescope array by the end of next year. But if they didn't get director's discretionary time, they couldn't get on ALMA until spring 2016. Worst-case scenario, they wouldn't get on ALMA until spring 2017, and that would set the project back by two years. That sort of delay was unthinkable.

Shep knew Pierre was talking with Heino's people directly, and that worried him. What were they telling him? Pierre was scheduled to see Heino and members of his circle next week at a conference in South Africa, and Shep didn't want them to have secret time. But Shep couldn't make it to South Africa, because he had so much else to do, not least of which was adding another important telescope to the array—the Large Millimeter Telescope in Mexico.

13

NATIONAL INSTITUTE OF ASTROPHYSICS, OPTICS AND ELECTRONICS
TONANTZINTLA, PUEBLA, MEXICO
APRIL 24, 2014

El Gran Telescopio Milimétrico, as it was known in its home country, was a giant silver bowl planted like a national flag on the summit of Sierra Negra, an extinct volcano on the eastern edge of the Mexican state of Puebla. At fifteen thousand feet, Sierra Negra stood a good five hundred feet taller than any mountain in the Lower 48, but it was overshadowed by 18,491 feet of glacier-capped splendor next door. The Spanish named the taller, more glamorous sister Pico de Orizaba. The Nahua people who lived here long before called the peak Citlaltepetl, or Star Mountain. In their language, Sierra Negra was Tliltépetl: Black Mountain. Getting to the summit of Star Mountain requires ice axes and ropes, so Black Mountain got the observatory.

The Mexican government built the Large Millimeter Telescope (LMT) in partnership with the University of Massachusetts Amherst. It was the largest scientific undertaking in Mexico, and with a movable dish fifty meters in diameter, it was the world's largest instrument of its kind. Construction began in 2000, and President Vicente Fox "inaugurated" it six years later. Engineers hastily installed a barely functional receiver they nicknamed Foxy so the president could go through with the ceremony. After the president's helicopter lifted off, the engineers got back to work, and five years later, the telescope reached minimum viable status. Before long, another helicopter landed on the gray andesite summit bearing the new president, Felipe Calderón Hinojosa. To avoid inaugurating the same telescope twice, they called President Calderón's ceremony a "visit of supervision."

Building a $50 million telescope in a troubled country posed challenges. There were rumors, hard to confirm, that early on someone stole dozens of precision-engineered panels meant to complete the surface of the dish, and that the police found them in a house where people were melting them down for scrap. The observatory's first director, Alfonso Serrano, now deceased, remains a figure of lore. People still tell stories about the raging parties he'd throw at a now-abandoned hacienda near the base of Sierra Negra; dignitaries would arrive from Mexico City in limos, mezcal would flow. The project's early history was so sketchy that astronomers spoke of it in tentative terms—*If the LMT is ever finished*... By 2014, though, the observatory was coming together. A new director, David Hughes, had professionalized the place. And the telescope was hugely attractive for the EHT. It was enormous, so it would be extremely sensitive. At fifteen thousand feet above sea level, it was higher than any telescope outside Chile. And its location filled an important gap in the EHT's coverage of the globe, providing a link between stations in North and South America.

But before it could join the EHT, the LMT needed upgrades. A receiver tuned for one-millimeter light. An additional mirror to send incoming light to that new receiver. The full high-speed signal-processing and recording rig that Jonathan and Laura and Rurik and others were developing back in Cambridge. A hydrogen-maser atomic clock. They didn't have the money for a permanent receiver, so Gopal Narayanan, the lead astronomer on the University of Massachusetts side of the LMT, was cobbling a temporary machine together out of spare parts. The Mexicans were building the new mirror. The signal-processing equipment wasn't finished. But one day in April 2014, it was time to give the LMT that atomic clock.

On the parklike campus of the National Institute of Astrophysics, Optics and Electronics, the Microsemi MHM 2010 Active Hydrogen Maser awaited. It was a bright, clear morning, and the air smelled like bougainvillea and street food.

Shep explained the mission to Betty Camacho, the office administrator, who, like office administrators everywhere, seemed to run the place. "It's a five-hundred-pound watch," he said. "And we're taking it into the apex cone room." The apex cone room was a vault-like concrete cylinder that rotated along with the telescope.

She smiled. "It's gonna be interesting to see how you get it there," she said in English. "How long is it staying?"

"Forever," Shep said. "We're going to turn the Earth into a telescope."

"What do you do with the hydrogen?" Betty asked.

"The hydrogen goes into a cavity and oscillates with very, very precise timing," Shep explained. The oscillations work as a "clock" so perfect that it should lose just one second every million years. "The problem is, if we bump it, it's just five hundred pounds of metal. And you don't really know if the maser is working properly unless you have another maser." After a moment, he completed the thought. "There's a lot of faith involved in science."

Betty showed Shep and an entourage of about a half dozen others to the high bay, where the maser sat in a white crate the size of a small refrigerator, shrink-wrapped, bolted to a wooden pallet, and covered with warning labels ("DANGEROUS GOODS IN MACHINERY"), export licenses, and Tip-N-Tell labels, which contain blue beads that spill out of their reservoir if the crate is tilted more than thirty degrees. Workers led by a friendly, authoritative man named Arak Olmos Tapia forklifted the crate into an air-ride truck, and it was time to go.

Usually it took two to three hours to drive from campus to the summit. With a maser in tow, there was no telling how long it would take. As Patrick Owings, the technician from Microsemi who Shep insisted come along for the installation, told Shep last night over dinner, "If you drop it more than about six inches, there's not much you can do. You have to send it back home." To which Shep said: "Yikes."

There was another complication. To prepare the maser for shipping, the manufacturer had to power it down and close off the vacuum chamber where the ions go. After six days in this state, it becomes increasingly likely that the maser will fall into a coma. The maser had been stuck in customs, and today was day seven. "Let's not dawdle," Shep told the crew.

"Okay, this is a little slower than I thought we were going to go," Shep said. On the highway outside Puebla, the caravan was moving at twenty miles an hour, flashers on.

Shep was sitting shotgun in a Chevy Suburban belonging to the Mexican government. Arak, in the air-ride truck, led the convoy, and his fear of potholes—*baches*—governed the pace. On his iPhone, Shep called Jonathan León-Tavares, a local astronomer who was riding in another truck, and asked him to gently nudge Arak into driving faster. "So, we have to get up to the top of the mountain tonight," Shep said. "I'm a little worried that if we go too slowly we'll be going up the mountain in the dark. We're behind you and we have the hazards on. Muchas gracias."

He hung up and said, "Okay, we're going to speed up to seventy kilometers per hour." Slowly, the caravan reached its new velocity, and the ride was smooth. "This is beautiful!" Shep said.

"I really don't like these clouds," Shep said. "Not because it's going to affect the astronomy, but because we could get stuck in a rainstorm on the way up. It could make the roads a mess."

Three hours outside Puebla, the mountain, Sierra Negra, had

come into view. Unlike its photogenic sister mountain, Pico de Orizaba, Sierra Negra was a homely mound of dark brown rock. From dozens of miles away, the Large Millimeter Telescope's gargantuan dish glinted in the sunlight—even, occasionally, now, despite the black rain cloud hugging the summit.

Everyone knew the drive to the summit would be rough. Roads in the state of Puebla are kept in decent shape, but they are also studded with Mexico's notoriously steep speed bumps, *topes*. The real concern was the dirt road that snaked to the top of the mountain—the highest road in North America.

The trucks left the main highway and turned onto a two-lane road, which soon grew rough and narrow. The atomic clock and its handlers rolled slowly through the cinder-block and terra-cotta villages of Atzitzintla and Texmalaquilla, past roadside shrines, wandering turkeys, idle dogs, enormous maguey cacti, and men guiding burros laden with bundles of wood. A herd of sheep surrounded the caravan for a moment. Sheep were a common hazard. Accoriding to LMT lore, a telescope operator once killed three lambs with his truck on the way down the mountain. He paid the villagers, kept the lambs, and hosted a feast.

When the pavement ran out, the road began to alternate between hexagonal cobbles and dirt. Cobbles and dirt, cobbles and dirt. And then, just dirt. At the sight of some horses, Shep started quietly singing "Tennessee Stud," the song Johnny Cash made famous. Then he gave a mini lecture on the history of astronomical observations to Igor Jiménez, a filmmaker from Mexico City whom Shep had hired to document the maser installation.

"After Einstein discovered relativity, they didn't know whether it was true or not," he told Igor. "Sir Arthur Eddington said, 'You know what we have to do is wait for an eclipse of the sun. And look at the stars very close to the limn. The starlight should be bent.' So they sent expeditions to Brazil and Africa. And, okay, I would say the weather is really closing in. This could be a problem. One doesn't plan to bring the clock up in a blizzard."

After clearing the guard shack at the base of the mountain, the trucks climbed a series of switchbacks through a forest of Montezuma pines. Rain came and then turned to snow. Above the tree line, with a few more switchbacks to go, the snow got heavy, and a thick fog surrounded the trucks.

"This is just amazing," Shep said. The valley below had vanished. "We've lost all visibility. We're just in the clouds." The trucks stopped. The weather at altitude is fickle, so the snow could either become a real problem—people had been snowed in at the summit for days—or it could stop at any moment. This time, the snow cleared almost as quickly as it appeared, and after several tense minutes, the trucks continued.

At the summit, the caravan passed through a large metal gate and parked in a gravel lot. Up close, the telescope's sheer size came into perspective. The silver bowl rested on a white mount, the whole apparatus the size of a Manhattan apartment building. The sky had cleared. The crew stood eye-level with the glaciers of neighboring Pico de Orizaba. At the edge of the parking lot, next to a work shed, someone had spray-painted an old wheelbarrow with the words, "Que me ves, pendejo?" *What are you looking at, asshole?*

Workers in navy blue coveralls explained the plan, and Jonathan León-Tavares translated for Shep. He pointed to an industrial crane as tall as the telescope. "*That's* the crane that's going to get the maser out?" Shep asked. Sudden exposure to altitude tends to make people delirious; Shep buckled over laughing. "I think that might be a little bit overkill!"

One of the drivers got back in the air-ride truck and drove down a concrete ramp leading to an industrial-size doorway. The workers took the roof off the cargo hold and wrapped the crated maser in green nylon straps and thick rope. A guy wearing a fanny pack holding what looked like an Xbox controller steered the crane into place. The workers hooked a cable onto the crane, everyone backed up, Shep swallowed hard, and with a few strokes of the controller, the crane slowly lifted the maser above the truck. "We have liftoff,"

Shep said. The crane operator set the maser down safely, and workers hand-trucked it into a large ground-floor room.

The next morning, after a night's rest at base camp in Ciudad Serdan, a village an hour's drive from the summit, the crew was back at the telescope, wondering how to get a four-hundred-pound atomic clock up a helical staircase.

The workers planned to hoist the maser up through the stairwell using a winch mounted two levels above. The maser was on level one and needed to get to level two. Level two was the ground floor of the soaring apex room. A metal staircase wound upward within the apex cone to a third level, where it landed on a slotted metal platform. The idea was to attach a winch to this platform, lower a chain to level one, hook it to the maser, and then winch it up to level two. Easy, except then the maser would be dangling in the middle of an

The Large Millimeter Telescope

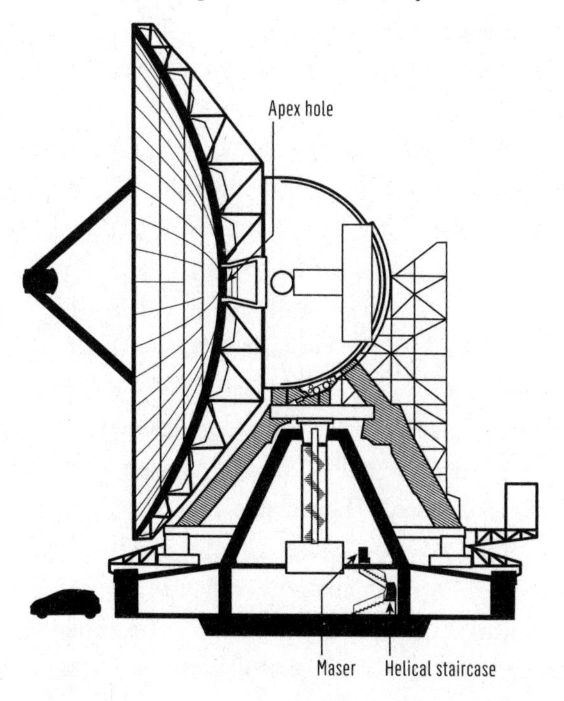

open stairwell; it would have to travel several feet to one side before it would have floor beneath it.

Translating the second stage of the plan was beyond the abilities of anyone in the room, so Shep watched in astonishment as the workers transformed into Sherpas. They hauled out aluminum ladders and roped them together as if to cross chasms in the Khumbu icefall. They cranked the maser upward until it was dangling a good fifty feet above the concrete floor of the bottom level. One guy put on a rappelling harness. They laid the makeshift aluminum ladder bridge across the wide, deep cavity through which the staircase ascended, and the worker wearing the harness, safety rope attached to the platform above, stepped out onto the lashed-ladder bridge. He pushed the maser toward another cable that the workers on the platform above had lowered into place. He clipped the second cable onto the maser, and then the workers overhead slowly gave the first cable slack, transferring the maser's weight onto the second cable. In this manner, Tarzan-style, they swung the maser, cable to cable, across the open stairwell. After about ten minutes, the maser was resting on the blue rubber-mat floor where it would remain indefinitely.

Uncrating the maser was a ceremony. To savor the moment, Shep and the others took turns removing bolts, like President Obama signing the Affordable Care Act with twenty-two pens. They saved the bolts in case they later needed to ship the maser, the prospect of which made Jason SooHoo laugh hysterically. They removed the walls and roof of the crate to reveal the maser itself, a black metal box the size of a small ATM.

Shep, Patrick, Jason, Jonathan León-Tavares, and Gisela Ortiz, a graduate student at the National Autonomous University of Mexico, set about bringing the machine to life. They ran extension cords and installed backup batteries and, gradually, opened a series of valves. They worked methodically, taking notes, workshopping each decision. At altitude, astronomers become hyperaware of their own fallibility. Thin air chokes the mind. Mistakes—"altitude moments"—are common.

As they counted each turn of their hex wrench, the workers descended on a stack of lumber. Shrieks of skill saws and pneumatic drills echoed throughout the conical concrete room. Soon they had erected a structure that looked like a rabbit hutch. This was the maser's new home.

"The next step is to start the hydrogen flowing," Patrick said to Shep. "Reach in there and give it one full turn. And if you turn it too far and it comes off in your hand, that's okay."

Shep knelt next to the maser, put his hand through a small opening near the bottom, and opened the flow. As he did, molecular hydrogen began to leak from the storage bottle into the discharge bulb, where an electric arc split the molecules into individual atoms. Those atoms would then pass through a state-selecting magnetic field and glide into a Teflon-coated quartz storage bulb contained within a tuned copper cavity. They would know the maser was working properly when the discharge bulb was glowing purple. "Okay," Shep said. "Now we wait."

They waited. They set about hooking the maser into the telescope's nervous system. They ran cables between the apex cone and the upstairs levels, three comparatively comfortable floors equipped with receivers, signal processors, servers, data recorders, monitors, and the computers from which operators steered and focused this monstrous instrument.

Hours passed. Early in the evening, the astronomers took turns kneeling beside the maser and peering underneath, as if inspecting the chassis on a sports car. The discharge bulb was glowing a cotton-candy shade of pinkish purple. This was a relief. The altitude was getting to everyone. "Ah, I have a headache," Shep said. "I feel like ten pounds of shit in a five-pound bag. As Mom used to say." They lifted the plywood rabbit-hutch enclosure and placed it over the maser. They laid a blanket on top, for warmth.

14

HARVARD-SMITHSONIAN CENTER FOR ASTROPHYSICS
CAMBRIDGE, MASSACHUSETTS
JUNE 11, 2014

The letter of intent that came out of the Event Horizon Telescope's 2012 kickoff meeting in Tucson promised that in time, they'd develop a memorandum of understanding that made the whole thing legitimate. Two years later, that memo remained unwritten. The addition of Heino's crew made the completion of that memo urgent. They needed to set rules and expectations for new members: what it takes to join, what's expected of you, what you get in return.

Shep didn't try all that hard to hide his bitterness about Heino's grant. As he saw things, by joining now, Heino's people were getting a decade of development work in which Shep and the rest of the early entrants had borne all the risk. How was that fair? It might have been a coping mechanism, but Shep thought they wouldn't be dealing with these issues if they weren't on the verge of something amazing. There was a reason people wanted in. He had taken to saying that the Event Horizon Telescope should be considered a major goal not just for astronomy or physics but for Science. He thought they could cause a historical disjuncture—before the first picture of a black hole, and after. That moment might only be a toilsome year or two away.

In the meantime, though, everyone had to agree on an org chart and the rest of the rules and bylaws that a finished memorandum would set forth. The next good opportunity to get everyone in a room to work through that memo would come in early June, when high-ranking people, telescope directors and others with weighty signatures, would already be in Cambridge for a conference celebrating the Submillimeter Array's tenth anniversary. In the months before that meeting, however, Shep was going to be busy shipping and installing masers and writing proposals and doing God knows what else, and he and the family were leaving for a trip to Israel later in June, so, in a departure, Shep asked for help. He sent the document he'd taken to calling the EHT Charter to a couple of institute directors and asked them to see what they could do with it. Later, he'd think of this as one of the moments when everything went wrong.

The next time he saw the charter it had accumulated many layers of legalese. That day at the CfA, in came the draft "Collaboration Agreement to construct and operate the Event Horizon Telescope," with its "NOW, THEREFORES" and its "supervisory and regulatory board, known as the EHTC Board"—after absorbing Heino's project, this Event Horizon Telescope–BlackHoleCam supergroup would be known as the Event Horizon Telescope Collaboration—and its EHT Management Group and that group's speaker, who reports "directly" to the board, which interacts with the Science Advisory Committee. Hardly anyone in the room had seen the draft, which was intended to replace the two-page letter of agreement they'd signed in 2012.

Shep was not a fan of this document. What did the director do but carry out the wishes of the board? Assuming he was named director—would he be?—could the board fire him? What's the point of the Science Advisory Committee? Did they plan to give the good stuff—the actual searching for the deep workings of nature in the finished, polished data—to a "council" of people with wood-paneled offices?

Everyone had questions about the charter. They spent two and a half hours talking about it, to the neglect of all other items on the day's agenda, including the most delicate question of all: if this newly formalized and expanded iteration of the EHT delivers the first picture of a black hole, who gets credit?

• • •

The night before they left for Israel, Shep sat up talking to Elissa, soaking in self-doubt. The National Science Foundation had emailed, saying they'd like to talk tomorrow about that $7 million grant proposal. He knew they'd already called one applicant to give them bad news. With these grants, the nos usually came before the yeses, and he hadn't been expecting an answer until probably September. Add to all this their lack of a formal collaboration agreement, delivery of which Shep thought might clinch their case. Throw in the pre-trip anxiety of flying a family of four to another continent. It's not hard to see why Shep was telling Elissa he knew they had blown it. Tomorrow, the National Science Foundation would inform him that the proposal that would fund the Event Horizon Telescope until first light had been declined.

But the next day, as Elissa and the kids scurried around the house loading up for the ride to the airport, Shep called the program officer at the National Science Foundation, and . . . well, what do you know? They hadn't blown it. The foundation had apparently overlooked (for now) the lack of a formal collaboration agreement, because they were recommending the $7 million proposal for full funding. Shep hung up the phone and fired a hastily written email to the collaboration announcing the news. Congratulatory replies washed in. In those brief, effusive notes, no one mentioned, though many knew, that they'd been competing with one of their own telescopes, CARMA, for the same pool of funds. They won, but CARMA lost, and as a result, it would probably shut down in less than a year.

PART THREE

FIREWALLS

15

Viewed a certain way, black holes are nature's way of telling us there's no hope: they will trap you, erase all record of your existence, and then vanish. This is too dark for most people. That's why theoretical physicists have spent decades reckoning with Stephen Hawking's discovery that black holes seem to destroy information. The goal has been to find a way for information to escape a black hole, which appears to be an inescapable prison on a deep, definitional level. In this pursuit, people have asked whether the information that falls into a black hole gets vacuumed into a new, baby universe. They've wondered if evaporating black holes left remnants, specks of primordial ash that encoded the history of everything that had ever fallen in. Neither suggestion has had the staying power of an idea, put forth in the early 1990s by Leonard Susskind, Lárus Thorlacius, and John Uglum, called black-hole complementarity.

Black-hole complementarity embraces the observer-dependent weirdness that repelled scientists in the 1930s, when J. Robert Oppenheimer and Hartland Snyder predicted that a person falling into a black hole would experience a completely different reality than someone watching from a distance. In quantum mechanics, light is both a wave and a particle; the form it takes depends on how it is observed. Susskind and crew proposed that something similar was true of black holes. In the Susskind-Thorlacius-Uglum scheme, a person falling into a black hole simply falls in, passing through the event horizon and plunging into the singularity, where he or she is crushed into some state of being beyond human understanding. But a distant observer watching someone else fall into a black hole sees that person become flattened and smeared across the event horizon. Both realities are equally true. They do not contradict each other, the argument goes, because it's impossible to produce an *experiment* that leads to a contradiction. That's because the only observer who could attest to falling through the event horizon intact is behind the event horizon, cut off from the rest of the universe.

Complementarity further holds that some strange surface just outside the event horizon stores information about the inner contents of the black hole. *All* the information—every single state of every single fundamental grain of spacetime, whatever those grains may be: strings, branes, loops. Physicists call these undefined states "degrees of freedom." We're familiar with two-dimensional surfaces that encode information about three-dimensional areas. We call them holograms. The idea that all the information about the interior of a black hole is engraved on a surface just outside the event horizon is therefore called the holographic principle.

It sounds absurd to say that the entire contents of some region—every single state of every single sub-sub-subatomic grain of being—can be encoded, like a hologram, on a two-dimensional boundary surrounding that space. But if you place faith in math and logic to reveal the undistorted foundation of things, this is where the argument leads. (Why can't a three-dimensional region contain *more* information than would fit on its two-dimensional boundary? Because if it did, it would collapse into a black hole.) Physicists have since extended the holographic principle to systems other than black holes. It might apply to the universe at large. It seems to be a fundamental aspect of nature—the way things work.

In 1997, Stephen Hawking bet John Preskill that black holes did, in fact, destroy information. By 2004, the information-rescuing holographic principle was widely accepted enough that Hawking conceded that bet. It seemed that the crisis had passed. And then, eight years later, it came roaring back.

Early in the spring of 2012, the late Joseph Polchinski, a theoretical physicist at the University of California, Santa Barbara, set out with his colleague Donald Marolf and two students, Ahmed Almheiri and James Sully, to fill in missing details in the standard picture of black-hole complementarity. They found, much to their surprise, reason to think black-hole complementarity didn't work—*couldn't* work.

Polchinski and his colleagues charged that the argument in favor of complementarity contained a subtle flaw with dramatic consequences. The flaw involves the phenomenon of quantum entanglement, a strange correlation that can develop between pairs of subatomic particles. "If particles were like dice," Polchinski wrote in *Scientific American*, "entangled particles would be two dice that always added to seven: if you roll the dice, and the first comes up as two, then the second will always come up as five, and so on. Similarly, when scientists measure the properties of one entangled particle, the measurement also determines the characteristics of its partner." Entanglement holds no matter how much distance separates the two particles—even if the two are separated by the event horizon of a black hole.

Entanglement plays an important role in black-hole evaporation. In quantum theory, a vacuum is never entirely empty. It always roils with the spontaneous creation and annihilation of pairs of virtual particles. Normally, those particles flash in and out of existence immediately. Near the event horizon of a black hole the story is different. There, gravity sometimes rips a pair of virtual particles apart as soon as they emerge from the vacuum. If one member of the pair falls into the black hole and the other escapes, the escaping particle becomes a "real" particle of Hawking radiation.

Complementarity holds that particles of Hawking radiation are entangled. Polchinski and his collaborators argued that if this is true, it leads to a contradiction. Again, Hawking radiation begins with the appearance of a pair of virtual particles. One falls in, one escapes. But those two particles are also entangled—and quantum entanglement is supposed to be "monogamous." An escaping Hawking particle can't be entangled with its virtual-particle nest-mate *and* particles that have already escaped the black hole. That means when Hawking radiation forms, the entanglement between the outgoing and ingoing particles must be broken. Breaking entanglement is a physical process that releases energy—enough energy, the Polchinski team argued, that the event horizon of a black hole should be a wall of fire that incinerates everything that hits it. In a sense, this "firewall" represents the edge of spacetime.

The firewall argument sent physicists into paroxysms. "I've never been so surprised," the theorist Raphael Bousso told the *New York Times*. "I don't know what to expect." Bousso said that the firewall paradox presented scientists with the "menu from hell." As reporter Dennis Overbye explained, "Either information can be lost after all; Einstein's principle of equivalence is wrong; or quantum field theory, which describes how elementary particles and forces interact, is wrong and needs fixing. Abandoning any one of these would be revolutionary or appalling or both."

Theorists proposed solutions to the firewall problem faster than anyone could read them. On January 22, 2014, Stephen Hawking joined the throng, posting a four-page technical paper online containing the words, "There are no black holes." This partial quotation made misleading headlines and spread across social media. "Stephen Hawking's Blunder on Black Holes Shows Danger of Listening to Scientists, Says [Michelle] Bachmann," read the headline of a humor piece published on the website of the *New Yorker*. Hawking was not, in fact, denying the existence of black holes: he was suggesting a new way of thinking about event horizons. Maybe the inviolable event horizon of old is more of an "apparent" horizon that traps light and matter only temporarily. Maybe it eventually vanishes, allowing the stuff inside to escape in incredibly garbled form. If correct, the black-hole information paradox and all the wild ideas that it spawned—firewalls included—would disappear. But despite his demigod status, Hawking's proposal was only one of many.

Steve Giddings, a professor at the University of California, Santa Barbara, had spent years trying to help information escape from a black hole. He thought he'd found the impenetrable puzzle's soft spot: locality.

The black-hole information paradox is often described as a two-sided conflict between general relativity and quantum mechanics. The truth is a little more complicated. There are actually three principles that come into conflict at a black-hole horizon: Einstein's equivalence principle, which is the basis of general relativity; unitarity, which requires that the equations of quantum mechanics work equally well in both directions; and locality. Locality is the most commonsense notion imaginable: everything exists in some place. Yet it's surprisingly hard to define locality with scientific rigor. A widely accepted definition is tied to the speed of light. If locality is a general condition of our universe, then the world is a bunch of particles bumping into one another, exchanging forces. Particles carry forces among particles—and nothing can travel faster than the speed of light, including force-carrying particles.

But we know that locality sometimes breaks down. Entangled quantum particles, for example, would influence one another instantaneously even if they were in different galaxies. The way Giddings saw it, quantum mechanics and certain aspects of general relativity have both proved nearly impossible to modify, and locality has its weaknesses. And after all, the whole reason black holes hide and destroy information is *because* of the principle of locality—nothing can travel faster than the speed of light, and therefore nothing can escape a black hole. If some sort of nonlocal effect could relay information from inside a black hole to the outside universe, all was well with the world.

His first approach was to ask whether, at some stage in the life of a black hole, a remnant composed of the stuff that had fallen in might somehow—through new, unknown physics—expand outward faster than the speed of light, breaking through the event horizon. Then he wondered whether there might be less violent possibilities. Is there some way that information could be imprinted on the energy that's already coming out in the form of Hawking radiation? He decided that there was one obvious way: if there were

The Event Horizon: Quantum Possibilities

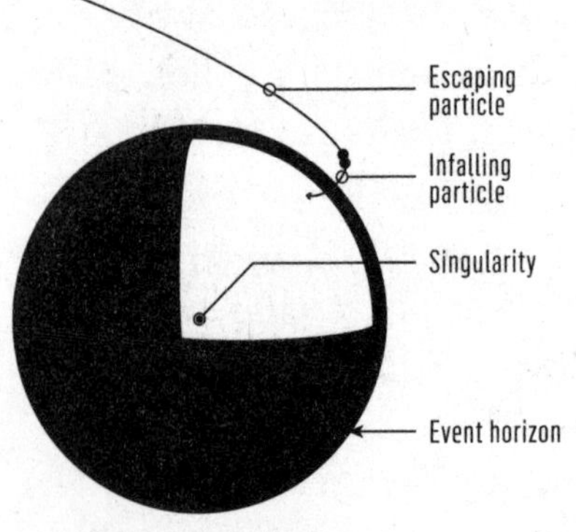

Non-Quantum View

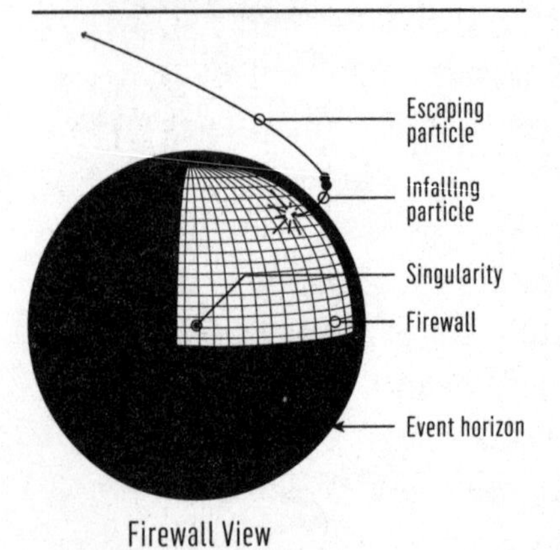

Firewall View

Escaping particle
Infalling particle
Singularity
Event horizon

Glistening View

gentle fluctuations in the gravitational field of the black hole; and if those fluctuations depended on the internal state of the black hole; and if those fluctuations made themselves felt outside the event horizon, influencing the Hawking radiation that gets carried away—then information trapped behind the event horizon had an escape route.

The idea, which he called nonviolent information transfer, was speculative, but there were so few ways out of this paradox that radical thinking was in order. And unlike some proposed solutions to the information paradox, it might be testable. To get information out of a black hole, these gravitational fluctuations would have to extend outside it. Giddings realized that if those fluctuations were large enough, they'd probably affect the propagation of light from around the black hole. Which meant that if they were real, there was a chance of seeing them.

Giddings first learned about the Event Horizon Telescope in 2012, when Dimitrios Psaltis visited Santa Barbara. In the subsequent months, he kept thinking about what these gravitational fluctuations might look like. By the spring of 2014, he had developed a mental picture. The fluctuations in the gravitational field could cause the light emitted by the black hole to gently scintillate. His favorite way to describe it was as a glistening. Information escaping the event horizon would glisten in the atmosphere surrounding the black hole. There was a chance—a small one, but a chance—that the Event Horizon Telescope could watch it happen.

16

MALDEN, MASSACHUSETTS
JULY 2014

Shep usually rose each morning around five, a respectable time for overachievers. These days, though, he was regularly waking up at three, his mind spontaneously seizing upon some element of a nightmarish two-and-a-half-page to-do list.

He'd typed up the list soon after returning from a not-exactly-relaxing family vacation to Israel during the outbreak of war with Hamas. On the day they visited the Holocaust History Museum in Jerusalem, sirens blared as they walked out the door. Everyone ran back inside and scrambled into an underground parking garage, from where they listened to the Iron Dome activate. Antirocket interceptors blasted off from mobile missile batteries. Midair explosions rumbled the garage.

Now Shep got to confront the imminent demise of CARMA. He had known that the Event Horizon Telescope was competing with CARMA for a fixed pool of grant money from the National Science Foundation, but there were political reasons that all they could do was cross their fingers and hope that both won. That didn't happen. Without the grant, CARMA would burn through what money it had left until next April, after which workers would crane the observatory's twenty-three big white dishes onto flatbed trucks and haul them down the mountain. CARMA's final expense would be restoring the land it had to a pristine natural state, as the U.S. Forest Service required.

On paper, the EHT could lose one site and still take a picture of a black hole. In practice, however, they usually lost one site per observation anyway because of bad weather. The EHT's strength—its efficient repurposing of existing telescopes—was only a strength if those telescopes continued to exist.

So they were going for it next year. What choice did they have? Computer simulations showed that for making images of Sagittarius A* and M87, CARMA was, if not essential, close to it. Shep kept hoping that some billionaire might Gulfstream in with $6 million a year and emboss his name on CARMA, but no one thought that was likely. It looked as if they had no choice but to pull the full Event Horizon Telescope array together between now and next March, somehow convince ALMA to participate, and, worst case, hope that new computer simulations would reveal that without CARMA, they were not, in fact, screwed.

So upon return from Israel, Shep typed up a deceptively short list of what had to be done before the spring 2015 observation. Every item on the list contained its own baby universe, each obeying its own laws of nature. The takeaway was that over the next several months, they had to install crates of equipment, some of which had not yet been built, at nearly every telescope they planned to use.

They'd figured out long ago that merely coordinating these telescopes wasn't enough to achieve their goals. They'd have to continue the long march toward higher bandwidth—faster electronics, bigger data packs. The goal for next year was to reach sixteen gigabits per second. To get there, they'd have to install a full kit of the latest VLBI equipment at every site: multiple Mark 6 data recorders, new digital back ends, new downconverters. The good news about the Mark 6s was that they existed. The bad news was that they only worked when someone who knew them inside out hovered nearby, ready to pounce if the machine stopped recording. The new digital back ends—those were still being designed. Jonathan and Rurik and Laura were writing the bitcode. Shep had $140,000 of computer chips sitting under his desk at CfA, which was funny given that his laptop was chained to his desk because he worked in an open building in the middle of Cambridge. The downconverters, which tuned the raw analog signal from the telescope's receiver to get it ready for processing, were also still being designed.

The South Pole Telescope needed GPS, a maser clock, test equipment, and a new receiver, which Dan Marrone and his team were building in Tucson. The LMT now had its maser, but Gopal Narayanan was still gathering spare parts for that telescope's makeshift Frankenreceiver.

Finally, there was the matter of getting time on ALMA. In August, Shep would fly to Chile and formally hand over the maser they had shipped down from Haystack that spring. While he was there, he'd meet with Pierre Cox, the observatory's director, and try to extract some assurance that ALMA would join them next spring. Since that cryptic phone call with Pierre back in February, Shep had received only noisy, ambiguous signals about ALMA's intent. But Shep was confident Pierre would understand the urgency of the situation. If CARMA couldn't be rescued, next spring might be the only chance to observe with a full, uncompromised, Earth-size telescope. His job was to convince Pierre that when next spring came, the EHT would be ready. If Pierre asked how much work they had to do before next spring, Shep would probably gloss over the details.

SANTIAGO, CHILE
AUGUST 2014

Pierre described his predicament to Shep over lunch at an Italian restaurant in Santiago. The NASA space probe New Horizons, sent to explore Pluto, was hurtling through the outer reaches of the solar system, less than a year away from its target, and its masters back on Earth weren't entirely sure where Pluto was. Sure, anyone could find Pluto with a telescope. But could anyone point a spacecraft traveling faster than thirty thousand miles an hour at just the right trajectory to ensure a rendezvous with the dwarf planet next July? No, they could not. So NASA sought help from the people at the Atacama Large Millimeter Array. ALMA scientists compared Pluto's progress against a quasar ten billion light-years away and measured the dwarf planet's parallax with twice the precision of previous measurements. NASA corrected the space probe's course. It was a feel-good moment all around, astronomers helping astronomers, until someone released the images that ALMA took of Pluto in the process of helping NASA. Then came the blowback. ALMA rules stated that those pictures shouldn't have been released because they were taken in an observing mode that wasn't officially cleared for public use. Pierre's point was that the mode Shep wanted to use—high-frequency VLBI—would also be an unsanctioned mode unavailable to the world's astronomers. And if he gave Shep telescope time outside the normal process, he'd never hear the end of it. "If I'm having problems with Pluto," Pierre said, "imagine the problems I'll have with the galactic center."

"Does the trouble scale with mass?" Shep asked. "If so, then we have a problem."

Shep was on the last leg of his Chilean mission. He'd gone to the ALMA high site, where he strapped on an oxygen-bottle backpack and visited the world's second-highest building (after a new station on the Chinese railway through Tibet), which housed the world's fastest single-purpose supercomputer, on which Shep's colleagues were performing surgery. He saw the Phasing Interface Cards they had installed, which would correlate signals from the hundreds of fiber-optic cables emerging from the floor. He hauled out his trusty handheld bomb-looking rubidium crystal and solved a mismatch between ALMA's resident crystal and the brand-new maser. He was relieved to find that the problem was not with his equipment: it was with ALMA's.

Now, over salmon with lemon sauce and a glass of wine, Shep could feel his special access to ALMA slipping away. His plan had been to project confidence, to argue that they had never embarrassed a telescope director in the past, and that if Pierre gave them a shot, they'd make him look good. When it became clear that confidence wasn't going to close the deal, Shep played the first of two remaining cards: He reminded Pierre that one of the telescopes they had always counted on, CARMA, was shutting down. This was their last chance to build their dream array.

Ah, Pierre countered, *but if you're so dependent on this one site, how robust can the Event Horizon Telescope be?*

Next card.

You'll recall, Pierre, that the G2 cloud is approaching the galactic center, and that gives us a once-an-eon opportunity to watch a black hole eat in real time.

But the G2 cloud is fizzling out, is it not?

This, alas, was true. The chances that G2 would fall directly into Sagittarius A*, setting off cosmic fireworks, had grown remote. Everyone had been watching, but no one had seen anything.

Every argument Shep made, Pierre jujitsued back on him with a smile. He never said "no." But Shep left Chile suspecting that his special access to the telescope that was essential for completing his life's work was, for reasons he didn't fully understand, close to being revoked.

17

HARVARD-SMITHSONIAN CENTER FOR ASTROPHYSICS CAMBRIDGE, MASSACHUSETTS SEPTEMBER 2014

Back from Chile, Shep turned to fund-raising, which wasn't the most natural move. The EHT had the money it needed to get to first light, and they had an overwhelming amount of work to do to get ready for next spring's observation. But the grant-writing never ends, and opportunities were appearing. Money was attracting money.

In stories in *New Scientist* and the *New York Times*, in the *PBS NewsHour* documentary they filmed when Shep was in Chile, in the press releases that the National Radio Astronomy Observatory and the other agencies involved sent out anytime the EHT crew achieved an incremental advance like installing the maser at ALMA, the experiment looked like a sure thing. Outsiders didn't know that Shep and Heino still couldn't agree on how to unite their groups—though they hoped to figure that out in November, at a big meeting in Waterloo, Ontario. They didn't know that CARMA, one of the telescopes the EHT scheme always counted on, was about to be dismantled. They didn't know that the team's access to ALMA was contingent upon the successful completion of unknown and constantly shifting milestones and requirements. People with money now wanted on board. Or, more precisely: they were interested in seeing a proposal.

The proposals were thick, polished documents, technical reports dense with nine-point type and plots and computer simulations. By September, Shep was working on several of these things, multimillion-dollar proposals to the Templeton Foundation, the Simons Foundation, the National Science Foundation. He was splitting the workweek between Haystack and CfA these days, so he'd carry his MacBook from his Haystack desk to his CfA desk with the occasional stop at his Hi-Rise Bread Company desk, pecking away at proposals, batting back at a bombardment of email. Much of the time, no one knew where he was. He seemed to like it that way.

Jonathan was baffled by Shep's fund-raising fixation. They had a trillion problems to solve before next spring. One of them, Jonathan thought, was that they weren't being aggressive enough in spending the money they already had. Jonathan thought of Shep as a Depression-era child, raised on scarcity, now, on the verge of success, stashing grant funds in the mattress. They needed to hire people—for example, the project manager they'd been talking about for more than a year, a professional organizer who would absorb the flood of logistical details in which Shep was drowning.

Jonathan spent his days working on his piece of the to-do list that Shep had night-sweated out earlier that summer. In a windowless lab on the ground floor of 160 Concord Avenue, down the hall from the postdoc room, Jonathan, Rurik, and Laura built and debugged new, faster versions of the back-end hardware that digitized and recorded light from the sky. They wrote code and tested Field-Programmable Gate Arrays, pushing them beyond industry-recommended limits, measuring path lengths between gates on the chip, sprinkling in delays—a nanosecond here, a nanosecond there—so that the signals that pulsed through the circuit every clock time were coordinated. They worked quickly. They had to get the South Pole Telescope's new digital back end to California by the end of September so it could ride with the Puckered Penguins down to the ice.

As their deadline approached, the days turned cool, and Cambridge swelled with returning students. It had been a year since Laura and Michael Johnson arrived, and the postdoc room at 160 Concord felt lived in. Conference posters hung on the walls, backpacks sat on the floor. On top of a particleboard bookcase stacked with electrical engineering textbooks was an espresso machine, which Jonathan bought because the BICEP-2 team down the hall had one. His thinking was that if you put in espresso, you get results. Now, he wasn't so sure.

BICEP-2 was a telescope at the Amundsen-Scott South Pole Station designed to study the cosmic microwave background, the cold bath of ancient light left over from the Big Bang. The cosmic microwave background fills the universe almost uniformly, with slight variations in density caused by quantum fluctuations in the first instants of cosmic history. These fluctuations formed the seeds of the future cosmos, motes of structure that over the eons snowballed into stars and galaxies. Back in March, Jonathan had muscled his way onto the crowded balcony in Phillips Auditorium to watch scientists from the BICEP-2 group announce the discovery of their careers. John Kovac, leader of the group, a hale, sandy-haired guy in his forties, told the assembled scientists and journalists that they had discovered faint swirling patterns, called b-modes, in the polarization of the cosmic microwave background. The theory of cosmic inflation, the leading model of the Big Bang, predicted that in the first unfathomably tiny fractions of a second of cosmic history, the universe ballooned from a trillionth the size of a proton to something the size of a softball and then expanded exponentially thereafter. That initial, impossible-sounding expansion involved such a ludicrous amount of energy that particles poured into existence from nothing. Among them were hypothetical particles called gravitons, which, in quantum theory, carry gravity. These gravitons caused the swirling patterns that the BICEP-2 telescope had apparently seen. If verified, the discovery would be the first direct evidence that the theory of cosmic inflation was correct. It would mark the first direct observation of primordial gravitational waves—ripples radiating through spacetime itself—and the first glimpse of quantum gravity and the graviton. It would also count as the first detection of Hawking radiation; the physicist Bill Unruh showed in the 1970s that extremely rapid expansion of space produces radiation just like Hawking radiation. That meant that the gravitons that produced the swirling patterns in the polarization of the cosmic microwave background were generated by the same process by which black holes evaporate.

At this news, people freaked out. "This is huge, as big as it gets," one cosmologist told the *New York Times*. In a YouTube video viewed nearly three million times, a cameraperson follows the Stanford University professor Chao-Lin Kuo as he knocks on the front door of his colleague Andrei Linde, one of the originators of the theory of cosmic inflation. Linde and his wife come to the door dressed for a morning of half-assed housework. "I have a surprise for you," Kuo tells them. Pretty soon, they're popping the cork on a bottle of Taittinger.

Six weeks after the Phillips Auditorium press conference, John Kovac made *Time*'s annual list of the one hundred most influential people in the world. But by then rumors going around said Kovac's team had misinterpreted their results. On May 12, *Science* published a piece with the headline "Blockbuster Big Bang Result May Fizzle, Rumor Suggests." The charge was that Kovac and company had made a mistake when they tried to subtract the distorting effects of galactic dust from their results. The telltale swirling patterns in BICEP-2's results might not have been ripples from the beginning of time—they could have been cosmic dirt floating in the foreground. After some protest, Kovac admitted that, yes, they might have made a mistake. But there was no way to know until September, when scientists working with an instrument called the Planck Satellite released results from the most detailed survey of the cosmic microwave background ever conducted.

When the Planck results came out, the verdict was harsh. It wasn't impossible that the BICEP-2 team had seen the imprint of inflation. But it *was* impossible to say that they had not, in fact, seen dust. The press declared BICEP-2's discovery dead with the same brio it had deployed back in the spring. As Dennis Overbye of the *New York Times* wrote, "Stardust got in their eyes."

Jonathan didn't think the BICEP-2 guys had done anything wrong, really. They got results; they announced them, with the caveat that theirs, like all scientific findings, must be duplicated; and then, when contradictory evidence came out, they admitted defeat. That was okay. They would try again. Kovac's crew would be flying down to the pole in a couple of months to upgrade the telescope to a more sensitive version. They called it BICEP-3.

But the BICEP-2 saga disturbed Shep. He saw it as a cautionary tale. In this trade, the business of peering into the unknown, when time came to share your discoveries, you had better make damn sure you weren't just seeing what you wanted to see.

Even when they had the full Earth-size telescope running, the black-hole shadow wasn't likely to sear itself unmissably into their data packs. They would have to fish it out of petabytes of digital noise. They would have to prove, to themselves and their peers, that they hadn't wished the black-hole shadow into existence—that they hadn't conjured an image out of a meaningless field of zeros and ones. So they'd have to install checks and balances. The Europeans, the Japanese, and the Americans would all analyze the same data independently, and each group would keep its findings quiet to avoid poisoning the minds of others. They could add additional layers of protection, too. For example, Shep had started talking to a group of machine vision experts in the MIT laboratory of Bill Freeman.

Freeman's lab in MIT's Computer Science and Artificial Intelligence Laboratory was a funhouse of extended perception. He and his students and postdocs had figured out how to amplify imperceptible motions in video footage—for example, the tiny rise and fall of a newborn's chest as it lay in its crib breathing; the oscillating skin tone (pale, flush, pale, flush) that every human with a pulse has but never sees—until they are visible. They called it the motion microscope. They invented a "visual microphone" that could reconstruct sound, including human speech, from the vibrations that sound induced in the leaves of a houseplant or a bag of chips. Shep wanted to know if they had any ideas for reconstructing images of black-hole shadows from sparse, noisy astronomical data.

Radio astronomers make images by measuring the characteristics of the light they collect and feeding that data to algorithms, which construct a picture of the object that emitted that light. For this purpose most radio astronomers use a decades-old computer algorithm called CLEAN, which was designed for single dishes. If the EHT were literally a telescope the size of Earth, making images from the light it collects would be straightforward, the results unambiguous and direct. But because the EHT was just a few specks of mirror on a rotating globe, an infinite number of possible images could explain any given set of EHT data. You could come up with a picture of frolicking unicorns that would fit the sparse samples collected by the EHT. How, then, could they be sure that the images they extracted from their data depicted what was really up there in the sky?

Katie Bouman, a grad student in Bill Freeman's lab, decided to work with the EHT. By the fall of 2014, when the BICEP-2 results were crumbling and Jonathan and Laura and Rurik were scrambling and Shep was nose-down in funding proposals, Katie was showing up a few days a week to work on new ways to pull pictures of black holes out of EHT data. She was developing a new algorithm called CHIRP, which stands for Continuous High-Resolution Image Reconstruction using Patch priors. CHIRP uses tiny pieces of existing images—these are the patch priors—like pieces of a jigsaw puzzle. Feed CHIRP data collected by the EHT, tell it the conditions under which that data was taken—which telescopes, in which configuration, in what weather, and so on—and the algorithm will sift through combinations of patches until it finds the most realistic image that fits that data.

So the postdoc room on Concord Avenue was full of life. Katie and a new grad student named Andrew Chael sat in one corner against the room's south wall and worked on dueling image-reconstruction algorithms. Michael Johnson pecked away at a study of the magnetic fields surrounding Sagittarius A*. Laura and Rurik ran tests and wrote code for firmware. The mood in the room was stressed but excited. They had big jobs to do, but they were discrete, narrowly defined, tractable: *get this equipment working by the end of September and get it in the mail.*

Shep, on the other hand, was increasingly occupied with the amorphous demands of diplomacy. When the National Science Foundation visited Haystack to check on the progress of the MSIP grant, they would be eager to see the collaboration agreement they'd been promised. Whether Shep would have an agreement to show them depended on the outcome of a meeting in Canada in November.

18

PERIMETER INSTITUTE FOR THEORETICAL PHYSICS WATERLOO, ONTARIO NOVEMBER 10, 2014

Maybe, Shep thought, *they should have held this meeting somewhere other than Waterloo. Could the negative connotations be any stronger? Why not have it in Pompeii?*

Waterloo is a small city an hour west of Toronto with a good university and a postmodern temple to fundamental science called the Perimeter Institute for Theoretical Physics. Mike Lazaridis, the co-founder of BlackBerry, founded the Perimeter Institute in 1999 with a gift of $100 million. The building looked like a sleeping Transformer, a jumble of matte black rectangles and iridescent windows. The floor plan could have been a puzzle for the institute's mathematicians to solve. The reason the EHT held its second biannual meeting here rather than Pompeii is that for a global collaboration, it was central. Also, Avery Broderick now worked here.

Just about everyone involved in the Event Horizon Telescope collaboration, and a couple dozen independent scientists, had come here for a week of talks and meetings. It was the first gathering of the entire collaboration since the founding meeting in Tucson two years earlier. As is often the case, the most important work wasn't listed on the official agenda. Between the science talks and technical sessions and conference dinners, they planned to hammer out that long-awaited collaboration agreement.

On the first morning, in the Theater of Ideas, Neil Turok, the director of the Perimeter Institute, gave welcoming remarks. He predicted that in the next two decades, "We are likely to see a major shift in theoretical physics, because there will be—there will need to be—new theoretical approaches that are consistent with nature's simplicity instead of predicting multiverses and chaos and other garbage." The Event Horizon Telescope, he explained, would be important in driving this progress. He said he'd been dropping the Event Horizon Telescope into public lectures and talks, building excitement. "Black holes are still the most paradoxical and bizarre objects in the universe," he said. "We desperately hope we'll learn something new about them."

Now Shep got up and spoke. Most of the talk was his usual boilerplate, but he took advantage of the new developments in the theoretical physics world and included the black-hole information paradox on the list of mysteries the EHT could explore. His slides declared that ALMA was still ready to join them in early 2015, although the likelihood of that happening diminished with every passing week. CARMA, as Shep delicately put it, was "under pressure." He was probably the only person who still thought there was a way to save CARMA—the people who worked there had already accepted its fate.

He didn't say that when they first started planning this meeting back in the spring, he thought there was no way they would not have the Event Horizon Telescope–BlackHoleCam merger figured out by now. He didn't mention his disbelief that this project, which had always worked so well on a handshake, had been thrown into bureaucratic hell. He didn't say that they were trying to organize a scientific experiment, not start a country, so why, in these collaboration talks, did it feel like they were drafting a constitution? Instead, he added a bullet point to the list of things he hoped would happen this week: "Organizational discussions."

• • •

The first day's agenda seemed designed to remind everyone why they did this work in the first place. Scientists studying black holes from noncompetitive angles described their findings.

Marta Volonteri of the Institut d'Astrophysique de Paris talked about the mysterious symbiosis between supermassive black holes and their host galaxies. They appeared to share a deep connection. In the late 1990s, astronomers studied some eighty galaxies and found that in all of them, the mass of the central black hole was correlated with the mass of the galaxy's central bulge. Why was that? Did the galaxy govern black-hole mass by regulating the amount of gas it's allowed to eat? Or did the black hole, despite its tiny relative size, rule the galaxy by blasting it with outflows that swept away gas, stopping galaxy growth and star formation? It was the duty of the people in this room to pursue questions like these.

Andrea Ghez from the University of California, Los Angeles, explained how her life had changed since her first big discoveries in the early 1990s, when she'd begun tracking stars orbiting Sagittarius A*. Around the same time Moore's law made Shep's work viable, an advance called adaptive optics revolutionized Ghez's. At the Keck telescopes on Mauna Kea, astronomers figured out how to shine lasers into Earth's upper atmosphere, creating artificial guide stars they used to measure the blurring effect of the atmosphere, and then subtract it by deforming the telescope's mirrors to compensate. The images they were taking now were at least ten times better, she said. They had now spent two decades tracking thousands of stars in the immediate environment of Sagittarius A*. She showed an animation of stars buzzing around the galactic center like fireflies, trailing glowing orbits behind them. She and her colleagues were gathering evidence to explain why Sagittarius A* is, against all apparent odds, surrounded by young stars—one of the reasons people once doubted that the galactic center held a black hole at all. And of course they were keeping an eye on G2, the gas cloud that the world had been waiting in vain for Sagittarius A* to maul into brilliant glowing pieces. G2 had survived its closest approach to the black hole intact, which, Ghez argued, meant it couldn't be a gas cloud at all. Instead, she hypothesized, it was a pair of stars being driven to merge by the presence of the black hole.

Coffee breaks had the feel of reunions. Distinguished guests brought the gravitas. James Bardeen, a soft-spoken man now in his mid-seventies, hung around between sessions, attracting admirers eager to meet the scientist whose forty-year-old throwaway calculations underpinned this entire enterprise. The atmosphere was humming, and because Shep was scheduled to give a public talk that night at the University of Waterloo, there was no chance that contentious collaboration talks would ruin the mood.

• • •

Tuesday, over lunch in a small meeting room, Shep and thirteen EHT principals talked about what they needed to do to get ready for the spring 2015 observation. The immediate focus was what observatories still needed what equipment, and how to get it there. The first question everyone had, though, was about ALMA. Were they in or out?

They were mostly out, but Shep couldn't bring himself to admit it. As they crowded around a long rectangular table, he said they had a chicken-and-egg problem. "If we were convinced we would be technically ready, I would go to the mat for it," he said. "I'd submit a director's discretionary proposal if I knew the technology was golden, and then I could press the case very, very hard. But all that evaporates if we can't answer the question, 'Will you be technically ready?' "

Geoff Crew spoke. "The only argument we have with them is that CARMA is going away."

"I spoke with Pierre in person about this," Shep said. "They are not impressed by this argument. They're a $1.5 billion instrument. 'Oh, you have a cut on your finger? Oh, that's too bad.' "

They set about deciding whether they would, in fact, be technically ready. Mike Hecht led the meeting. "Presumably everyone has a maser that needs a maser," he said. "Anyone not have a maser? All right, good start."

Remo Tilanus had found himself becoming Mike Hecht's European equivalent—a volunteer project manager, a professional adult in the room. He joined Hecht in running down the list. "Basically all stations will get two Mark 6 recorders," Remo said, "and CARMA will get an additional one for the reference station. That's the baseline plan."

"And are they ordered, and being shipped where?" Mike Hecht asked.

"They should be at Haystack now," Remo answered.

Mike noticed Shep and Jonathan whispering at their corner of the table. "Can we have one meeting, please? Thank you."

They continued down the list. What's the status of the quarter million dollars of helium-filled hard drives? What about the digital back ends? Could every site get oscilloquartz crystals and redundant sets of cables and connectors? The meeting devolved into a polite group argument about what equipment was desirable and what was necessary, and with no clear conclusion, they ran out of time. The scheduled conference resumed right after lunch.

• • •

On Wednesday, the dispatches from slightly tangential fields continued. Samir Mathur explained how "fuzzballs" could solve the black-hole information paradox: a new form of matter might at some point exit a black hole's event horizon, carrying with it the information that had been concealed within, like a neutron star emerging from a cocoon. That afternoon, Gabriela González, spokesperson for the Laser Interferometer Gravitational-Wave Observatory, or LIGO, gave an update on the hunt for gravitational waves. LIGO, a pair of miles-long installations that would use lasers to measure the minuscule deformation of spacetime caused by distant black-hole mergers and other unfathomably violent events, would switch on late next year. If it was successful, they'd soon directly detect gravitational waves, proving yet another of Einstein's predictions and opening a new observational window on the universe.

On Thursday, the focus switched to logistics. The astronomers gave updates on the vast deployment of matériel that had to happen in the next few months. Gopal Narayanan talked about the status of the receiver he was building from scraps for the Large Millimeter Telescope. Dan Marrone described the forbidding logistics he'd face next month when he traveled to Antarctica to install a one-millimeter-wavelength receiver at the South Pole Telescope. Laura Vertatschitsch explained how she'd written the firmware code that would power the digital back ends that every EHT site would use for next spring's observation.

And about next spring's observation: in his talk on the state of the effort to turn ALMA into a phased array, Mike Hecht delivered the news that no one else would, news that Shep still couldn't quite accept—ALMA wasn't going to join the EHT next spring. This in no way diminished the urgency of the 2015 observation. They had to get all the sites working together and prove to ALMA that they were ready for the big full-array imaging run. But they weren't going to get the shadow image next spring. They'd do it in 2016. The schedule that Shep had set years earlier would slide by a year, and that wasn't the worst thing in the world.

• • •

Thursday night was movie night in the Theater of Ideas. Avery had wanted to show *Interstellar*, the Matthew McConaughey–Jessica Chastain space opera based on the ideas of famed black-hole researcher Kip Thorne, which involved a hyperrealistic simulation of a supermassive black hole named Gargantua. The movie had just come out, though, and astronomers didn't have the pull to get a private screening. Instead, they showed *Event Horizon*, the 1997 film in which a spaceship powered by an artificial black hole accidentally opens a portal to hell.

The choice of film made a decent metaphor for the meeting happening concurrently upstairs. In a closed-door session, Shep and Heino and a dozen or so others debated their organizational structure for hours. The philosophical question the astronomers faced was, What sort of organization was the EHT going to be? Would it be an observatory, a big telescope standing by that any astronomer could apply for time on? Would it be an experiment, a coalition of scientists banding together to achieve a specific, well-defined goal, in this case capturing the first image of a black hole? Or would it be something in between: a hybrid, with bureaucratic eccentricities to match? The practical question they faced was one they couldn't say aloud: how do I make sure my name gets on the Nobel Prize? Now, while the organization was still being formed, was the time to jockey for position. So as Sam Neill and Laurence Fishburne battled evil forces from another dimension downstairs, the astronomers negotiated in the Black Hole Bistro, a restaurant on the Perimeter Institute's second floor. The place was closed for the night, so they were free to pace the whole dining room as they rubbed their temples and sighed heavily and shouted and cursed.

At the end of the night, Shep and Geoff Bower walked into the bar at the Delta hotel across the street glowing and exhausted, as if they'd just finished a half marathon. They ordered a couple of scotches. They had reached a preliminary deal. Shep had a photo on his iPhone of the astronomers posing for a group shot in the empty Black Hole Bistro. Avery had emailed it to him, with the subject line "A collaboration is born!"

• • •

The next day, before the farewell lunch, they unveiled the details of the collaboration. Colin Lonsdale did the talking.

He began in the cautiously upbeat mode of someone who knows he's about to disappoint and confuse a large fraction of his audience. This has been "an absolutely terrific conference," he said. "Just a constant stream of absolutely wonderful talks. It's all been overshadowed by a minor tragedy. There have been fifteen to eighteen people who have not had the opportunity to see these talks. But the sacrifice has not been in vain. We have actually made major progress toward a robust and workable collaboration."

Before getting into the details, he wanted to make sure everyone in the room understood how hard this had been. "This is a far-flung collaboration," he said, "global in nature, with engineers and observational astronomers, theory groups, and many different organizations with different institutional cultures. It involves a tremendously complex observing effort and engineering challenge, and technology development in a very aggressive way. At the same time, the collaboration is growing rapidly. This is world-class science that the EHT is going after. And then just to spice it up a little bit, the collaboration is on the threshold of revolutionary observations that are going to lead to major, major results. All of this is wonderful, but it requires some level of organization to keep things coordinated."

He read a summary of the results of the week's side meetings. There will be one Event Horizon Telescope; BlackHoleCam would be absorbed into it. At the top of this new organization—the Event Horizon Telescope Collaboration—was a board composed of representatives from "stakeholders," those who have "formally committed significant resources to directly advance" the project. The board will appoint a director, who will report to the board. An entity called the Science Council "will develop the prioritized list of science objectives." The collaboration will hire a project manager and a project scientist. And then there will be technical working groups, science working groups, and so on.

Shep started having second thoughts as soon as they left Waterloo. Everyone likely to be on the interim board—mostly the same people who had signed the first EHT letter of intent back in 2012—said they'd elect him director, but what was the director in this scheme but a hired hand who reports to the board? Besides, he didn't really want to be the director of an observatory—he wanted to be the lead scientist on a historic experiment. Yet the structure they'd worked up in Waterloo cleaved off the prestige work and handed it to the science council. And the plan included little assurance that he'd remain in charge. There might be a coup. He might not be elected director. Or the board might fire him. He could imagine a plausible scenario in which the Europeans seized control of the project he'd spent his entire career building.

• • •

They didn't teach you to navigate treacherous political waters in grad school. On a family vacation in New York over Christmas break, Shep walked into a bookstore and bought three titles: *Getting to Yes*, *The Power of a Positive No*, and *How to Have That Difficult Conversation*. He wondered what the clerk must have thought. *Good afternoon, sir, I see you're trying to build an Earth-size virtual telescope.*

Two months passed after Waterloo before the interim board invitations went out. When they did, many of them got kicked up to people's bosses. The institutionalization of the Event Horizon Telescope accelerated, and Shep couldn't overcome the nauseating sense that this thing was becoming too big for him to control.

19

As the political machinations continued, preparations for the spring 2015 observation accelerated. In December 2014, Remo Tilanus flew to Haystack for a three-week visit, and while he was there, he and Mike Hecht decided that while everyone else argued about the org chart, they were going to make this spring campaign happen. The plan for years had been to hire a project manager, but that was a big expense, and Shep had never felt the project could afford it. As a stopgap measure, Mike and Remo's transatlantic alliance would share those duties.

That same month, Dan Marrone flew to Antarctica to get the South Pole Telescope ready. When he arrived, two of the thirteen crates of equipment he'd shipped down were waiting for him. He'd been planning this upgrade—adding a hydrogen maser, a new receiver, and all the rest of the requisite VLBI equipment, thus enabling the South Pole Telescope to join the EHT—for years. But he'd only been here once before, and on that trip he'd spent only a couple of hours inside the cramped cabin where he now had to install a hand-built supercooled telescope receiver the size of a small motorcycle. Turns out the dimensions they had to work with were different than they understood. Which meant they had to place the mirrors that relayed incoming light to the receiver in a slightly different configuration than expected. Which meant they had to mount every downstream component in a slightly different configuration than expected. They spent hours measuring the distance between things that were six feet apart to within a millimeter and then making corrections. Also, the place was under construction. The glycol loop compressor that heats the buildings and cools the telescope's cryostats was being rebuilt, so for the first month and a half of Dan's two-month mission, the ambience was dominated by the flashes and acrid smoke of acetylene torches.

By the time all thirteen crates arrived, Dan and his crew were running seriously behind. They had to install all their equipment, test it, and then get off the pole before the weather turned. Seven days a week, including Christmas and New Year's, they'd start around eight in the morning, break for lunch and dinner, and then work until midnight. It was draining, but what was there to do but work? Dan was surprised to find the pole relaxing, even meditative. Down here he had one job, round-the-clock sunshine, and no distractions. He had a wife and two kids back in Tucson, including an infant daughter, and the time apart was a strain. He talked to his family almost every day on the satellite phone, and managed to exchange some pictures over the atrocious Internet connection. But he had virtually no email and no social expectations. He worked fifty days straight and never got a headache.

So they installed the receiver and the maser, cut a hole in the roof of the building that housed the telescope, and ran hundreds of feet of cable. They set up a system in which, each year, when it was time to switch the South Pole Telescope into black-hole-hunting mode—it spent the rest of the year measuring the cosmic microwave background—someone would place a small aluminum mirror in a special backpack and climb a scary fifteen-foot ladder to the roof in total darkness and negative sixty-degree temperatures. He or she would drop the mirror onto its mount on the roof; the mirror would relay the light collected by the telescope's main, ten-meter mirror into the VLBI receiver. And when it was time to switch the telescope back into its normal operating mode, someone would climb up the ladder and do the process in reverse.

On January 16, 2015, Dan climbed the ladder and dropped in the special mirror himself. The telescope's operators pointed the dish at the moon and then at Sagittarius B2, the huge molecular cloud in the galactic center. It was just a test, but the receiver worked perfectly.

And then the next day, they started disconnecting the receiver. Dan still needed to add another, higher-frequency, mode to the receiver, and if ALMA wasn't going to participate in the big Event Horizon Telescope run three months from now—if the long-awaited big imaging run wasn't happening this spring—he didn't see any point in leaving it at the South Pole for another year when he could bring it home and work on it in his own lab in Arizona. He'd have to redo everything next year anyway because the station was scheduled to be renovated in the coming months. So they shipped the receiver back to Tucson and made it official that the South Pole Telescope, like ALMA, would be sitting out this year's observation.

HAYSTACK OBSERVATORY
FEBRUARY 3, 2015

As Shep and Jonathan and the rest of the Boston-area crew worked to ship equipment out for the spring campaign, near-weekly blizzards lashed the region. The snow came in two-foot slabs, one after another, layer upon barely shoveled layer. The city of Boston postponed the day's victory parade for the New England Patriots, who had beaten the Seattle Seahawks Sunday in the Super Bowl.

Shep's kids hadn't been to school for a week. They were old enough that Elissa had taken her career out of maintenance mode: she was saying yes to the things she used to say no to, going on the road, preparing to apply for a long-delayed promotion to associate professor. This week, though, she'd been home with the kids the whole time, watching the snow pile up, going stir-crazy. Shep kept shoveling out the driveway and making the treacherous drive to Haystack.

The countryside around Haystack looked like a maple syrup ad. Mounds of snow the size of bulldozers dotted the Haystack parking lot. Everyone was late, but no one took a snow day. Gopal drove in from Amherst. Jonathan was here with his two young kids. Vincent Fish, Geoff Crew, Mike Hecht, Jason SooHoo, and a new postdoc named Andre were all here.

Under fluorescent lights in an engineering room, the crew pondered the configuration of the digital back ends and Mark 6 recorders they were about to ship to Mexico for installation at the Large Millimeter Telescope. The space was a semi-orderly jumble of cardboard shipping boxes and workbenches and racks and rolling carts holding Tektronix oscillators and Agilent network analyzers and spools of cable of varying gauges and colors. At the far end of the room, the equipment under scrutiny had been placed on low tables, wired up, and powered on. Fiber-optic cables hung from the wall behind the equipment.

Laura Vertatschitsch had pretty much single-handedly built the digital back ends they were testing, but she was in Cambridge today, and Shep was confused, so someone emailed her and asked her to call. Jason had a warning for Shep: Laura, who grew up in Seattle, was a little upset about the Super Bowl. "Don't mention the game," Jason said.

The phone rang. Vincent answered, and then handed it to Shep.

"Hello, Laura," Shep said. "Are you mourning your beloved Seahawks?"

Jonathan and Jason recoiled, smiling mischievously.

"I bet you are," Shep said. "With the proper medication and counseling we'll get through this. Listen, we're looking at the back end. Andre is running a script, and we thought that was supposed to go through and perform all the calibrations. . . ."

Shep hung up the phone and turned to Jason and Vincent. The issue was that they were using the recorders in a different mode than they had in the past—one that they hadn't tested, and one that Shep was concerned about switching to immediately before shipping the equipment to Mexico.

"We can figure it out," Jason said.

"I understand we can figure it out," Shep said. "I just don't like doing things for the first time the day before we ship it out. I also don't like planning to do something in the field that we've never done before."

"Two points," Vincent said to Shep. "This is not the mode that's been done before. And some development and testing needs to be done. But regardless of whether we use one Mark 6 in the field or two, if we have a site where the second Mark 6 fails, we need to have this mode of operation at our fingertips. It's either our primary operating mode or backup operating mode."

"I'm fine with it being either as long as we test it," Shep said.

"We also have a meeting with Geoff in six minutes," Vincent said.

"You're kidding me."

• • •

On the other side of the building, in Haystack's main conference room, Shep and Vincent and Geoff Crew met Mike Hecht and Lynn Matthews, a Haystack astronomer who devoted much of her time to the ALMA Phasing Project. Last month, Geoff and Lynn had gone to Chile for the first in a series of commissioning tests; if the tests were successful, ALMA would formally accept the phasing project, which meant they would soon be able to use it. But the phasing system had a perplexing bug, a fault in the system that corrects the tiny delays in the time it takes for signals to travel from ALMA's sixty-six antennas to the supercomputer that correlates them. They had a workaround, but it wasn't pretty. The point of this meeting was to decide whether to stick with the workaround, or hunt for the root of the problem. The second option could delay the completion of the phasing project, and thus the first Event Horizon Telescope observations with ALMA, for longer than anyone wanted to contemplate.

They took seats around a budget executive boardroom table consisting of seven smaller rectangular tables pushed together. Shep, Mike, and Vincent sat opposite Geoff and Lynn, deposition-style. Shep, Mike, and Geoff all leaned back in their chairs and folded their palms on top of their heads.

"So we've identified the problem with residual delays," Shep said. "We need someone who understands all of the solutions. Get everything on the table. It would be really disastrous to miss another thing like this."

"Part of the problem is there aren't that many people left who know the whole system," Geoff said. "I'm not sure there is an expert we can take this to."

The feeling was that this subtle, unexpected technical gremlin was no one's fault, really. During the long, tedious process of vetting the design for the ALMA phasing system, a few people halfway realized that the residual delay system was something they'd eventually have to contend with, but no one thought it would be a problem. And when they first started testing the phasing system, it wasn't a problem. Then they switched to the higher frequencies that the EHT needed to work with, and the tiny errors added up. Meanwhile, most of the engineers who built the system had left for other jobs.

"My feeling is we should write this up," Shep said. "Memo A is, What is the exact problem? Then: What knobs do we have to turn? Another memo!"

"Part of me wants to say that a memo well written enough for distribution is not a quick thing," Geoff said. "If we're going to say we screwed up, it has to be a well-written memo."

"If we don't understand this thing really well, we're not going to be able to fix it," Shep said.

Mike turned to Geoff. "Is this the time to call in the cavalry, or leave it to you guys?"

"Frankly, I think us doing it is the only way to finish the project by August, which leaves several months of politicking to figure out what's happening in March 2016."

Since the Waterloo meeting, when Mike Hecht made everyone face the reality that ALMA wouldn't be joining them for the next campaign, they had come to accept a one-year delay. But a one-year delay was starting to sound like the best-case scenario. Doing major surgery on the phasing system could push back the first big full-array EHT campaign indefinitely.

"Let me just say something directly to Mike," Shep said. "The real problem is that this is the last year of our grant. If we ask for an extension, then we have to clearly explain the problem. This is our reputation. We get grants because of our reputation."

Everyone was quiet for a moment. Back in 2011, the National Science Foundation gave them a grant to phase up ALMA, separate from the grants the Event Horizon Telescope used for everything else. That grant was nearly up. If they weren't going to finish by August, they'd have to formally request more money and more time.

"We're not going to conclude on our current timetable," Shep said after an uncomfortable pause. "For my money—that's the wrong way to put it."

"For the NSF's money," Vincent said.

"For the NSF's money," Shep said.

They spent a few minutes discussing their options, none of which were appealing. They were scheduled to run tests on ALMA in the next few weeks. If they paused to write a big diagnostic memo, they'd miss those tests. The next opportunity to do those tests would be months later, so taking a few extra weeks to solve this problem would translate into a delay of at least a year.

"March is open season for testing," Geoff said. "Taking time now means not testing in March, which means we're not going to finish commissioning in 2015. And that's fine if we're making that choice consciously."

No one was ready to make that choice. For another several minutes, they talked, at first rehashing the same arguments, eventually following a tangent about whether ALMA knew the precise position of their telescope pads on Earth's surface or only their location relative to one another.

"Can I rewind to what we do next?" Mike said.

"Now I beat my head against it," Geoff said.

Shep took off his glasses, rubbed his temples, and lifted a single-page data plot from the test run to eye level. His desire to solve this problem with a sudden moment of insight could have burned a hole in the paper. After a while, he set the plot down on the table. "Lots of hard work has been done," he said. "We need to honor that. Separately, Mike and I have to think about the NSF and external funding. I liberate you from that."

Geoff asked if they were concluding that they wouldn't finish the phasing project by August.

"We're not concluding that," Shep said.

Mike asked how far in advance they had to tell ALMA whether they planned to do those tests in March. About a month in advance, Geoff and Lynn said almost in unison.

"So we have maybe a week?" Mike said.

If they skipped doing tests with ALMA next month, Geoff asked, would they even be able to get telescope time for the full EHT array to run in 2016? Or were they looking at an even longer delay?

Shep rubbed his head and closed his eyes. "You know, I feel like I'm going to become a diamond because of the pressure."

"Pierre Cox was clear about this at the Bologna meeting," Vincent said. Last month, Shep and Vincent had flown to Bologna for a meeting of the European VLBI community, where Pierre stood in front of the crowd and said that the rules for getting on ALMA were inflexible. "We had to have our ducks in a row by December, apply by March, get on Cycle Four, which is late 2016, so that means we're doing our first observations in March 2017."

"Oh God," Shep said.

20

LARGE MILLIMETER TELESCOPE
SIERRA NEGRA, PUEBLA, MEXICO
MARCH 19, 2015

An hour before sunset on the first night of the 2015 observing campaign, Laura Vertatschitsch stood inside the cylindrical metal tunnel leading to the truck-size hole at the center of the Large Millimeter Telescope's fifty-meter dish. She was joined by Lindy Blackburn, a new EHT postdoc, and David Sánchez, the LMT's operator. A few minutes ago, workers had removed the blue tarp that covers the hole—known as the vertex hole or apex hole, depending on whom you ask—when the telescope is not in use. The three were dressed for cold: hoodies, stocking caps. Laura peered over the edge, where a terrifying ski slope of polished nickel led to doom on the volcanic rubble of the summit. The blue sky was turning orange. Towering clouds floated by. A wispy haze blew off the mountain.

"Look at this," Laura said. "Who gets this kind of job?"

David ran downstairs to the control room to give everyone a ride. Soon the telescope silently began to rotate to the left and the sunset-drenched Mexican countryside fifteen thousand feet below panned by. "Yeehaw!" Laura said. "We are moving! We are in the telescope and we are moving. Oh, this is spectacular."

Below the mountain, toward Veracruz, a storm was forming. The sunset was a deepening orange.

Laura sighed. "We're sitting here in the heavens."

The telescope came to a soft, silent stop. "Ride's over," Lindy said.

• • •

Light pinballs through the LMT. The giant parabolic dish (properly known as the primary reflector) focuses light from space (the sky signal) onto a two-and-a-half-foot mirror (M2) mounted on booms that extend from the main dish like a selfie stick. From M2, light enters the apex hole and travels through the tunnel where Laura and Lindy and David were hanging out. At the back end of this tunnel, a piece of machined aluminum called the tertiary mirror (M3) sends the sky signal to a fourth mirror. This mirror—M4—is one of the upgrades the EHT scientists and their local collaborators installed to prepare for this year's campaign.

M4 beams the sky signal through a hole someone sawed into a panel of Plexiglas on down to M5, a square of aluminum about the size of a textbook. From here, it's one more bounce and a few inches of travel before the light enters the interim receiver that Gopal Narayanan cobbled together just in time for the observing run.

As Lindy and Laura climbed out of the vertex-hole tunnel, hopping down from the mirror cabinet onto the tiled receiver-room floor, Gopal and his student, Aleks Popstefanija, were calibrating the system. Gopal was a lean, alert man in his fifties, with jet-black hair, a goatee, and rectangular glasses. He was dressed for the mountain: forest-green fleece over a blue flannel shirt, jeans, hiking shoes. He placed a Styrofoam cooler full of liquid nitrogen on a stack of boxes in front of M5, and Aleks, a recent University of Massachusetts Amherst graduate, wrote down numbers in a notebook.

The interim receiver was a collection of gadgets bolted to a white rectangular tabletop. After leaving M5, light passes through a beam splitter and a thin layer of protective white foam and on into a Plexiglas box, where some fraction of that light enters the tiny mouth of a gold-plated feedhorn. Out of the small end of this horn emerges the cosmic light worth keeping. This final slice of light enters a mixer, a block of metal into which machinists, peering through microscopes, have carved a series of canals and slits. Here the sky signal is combined with a pure tone piped up from the hydrogen maser several floors below. The difference between the sky signal and the maser signal is called the intermediate frequency, and it is much easier for electronics to handle than the high-frequency sky signal.

The intermediate frequency travels through another microscopic channel before striking the superconductor-insulator-superconductor mixer, which converts it into an electrical signal encoding information about the phase and amplitude of the light. The loud, rhythmic chirping noise that fills the receiver room is the sound of compressors pumping liquid helium to cool the mixer to four degrees above absolute zero. At this temperature, the mixer is so sensitive that its users must account for the effects of quantum noise.

A black hole shines because the atoms around it emit photons. Those photons travel for twenty-six thousand years before striking a nickel-plated dish the size of the infield of a major-league baseball diamond. They ping-pong through a maze of aluminum mirrors before being converted into electricity inside a telescope receiver. The jittering atoms from which the receiver is composed generate a roar of thermal noise a hundred thousand times more powerful than the signal from the sky. There are lots of opportunities for error in this unlikely series of events and conversions. Which is why the testing and calibration that precedes every observing run is a lengthy sacred ritual. Right now, Gopal and Aleks were measuring the "temperature" of the infrared light emitted by the cooler of liquid nitrogen so they could determine the baseline noise of the receiver. They would subtract that baseline from all the measurements the receiver recorded. As far as they could tell, everything looked good.

• • •

The electrical signal encoding light from the black hole leaves the receiver and travels through coaxial cables two floors down to the back-end room, where the signals are digitized and recorded. Shep and Lindy had installed the racks of electronics that perform this task three weeks ago.

It was quiet in the back-end room. The space was a jumble of server racks and Pelican cases and work tables strewn with cable connectors and attenuators and other electronics-shop gear. Jonathan León-Tavares, the Mexican scientist who had joined Shep for the installation of the maser here last spring, flipped through technical diagrams. Laura and Lindy, now down from the receiver room, rummaged around for Ethernet cables they could use to control the Mark 6s' data recorders that nearly every site in the EHT would be using this year. Shep sat at a worktable, wearing a knit beanie he had borrowed from his kids, somberly focused on his laptop.

The weather gods had been sending mixed signals. A week before the campaign, a snowstorm battered Sierra Negra, encrusting the telescope in ice. A week before that, there was a big thaw, and melting ice on Pico de Orizaba revealed the remains of two climbers who had died in an avalanche in 1959.

Gopal walked into the room, and he and Shep talked about what they needed to get done before Gopal left the observatory in approximately thirty-six hours to join his wife back in Massachusetts for some medical tests. "I'm here all night tonight and all night tomorrow night," Gopal said.

• • •

Next door, in the control room, David Sánchez was trying to focus the telescope. It was a little after 8 P.M. In about an hour, M87 and a couple of bright quasars would come into view, creating the first opportunity to make sure the LMT and the rest of the telescopes in the EHT array were seeing the same things. But the weather might prevent that from happening.

Here in Mexico, the situation was mixed. The tau was about 0.5, which was bad, but on this mountain, the sky cleared late at night reliably enough that local observers spoke of the "nasty 4"—4 P.M.—and the "beautiful 4"—4 A.M. The other sites were sending conflicting signals. On the Google document that Lindy and Laura set up to record local weather reports, the SMA crew wrote that the weather was "Good," while someone next door at the JCMT said the weather was "BAD."

Before they could do anything, David had to focus the telescope, and to do that, he needed a planet to look at. Jupiter was too big and too bright for fine-tuning a telescope like this, but Saturn wouldn't be up till 2 A.M., and Venus was already down, so Jupiter would have to work.

Around 8:30 P.M., Shep stepped into the control room and walked over to David, who was sitting at a large bank of computer monitors. One of the monitors displayed the latest weather data.

"The blue level," David explained, "is the level of crappiness"—the clouds, rain, and other forms of water vapor that could interfere with the observation.

"Are you telling me that for the foreseeable future the weather is going to be crap?" Shep asked.

"In a manner of speaking."

"Are you able to focus this dish?"

"If I get good weather, yes." David had managed to detect Jupiter—which wasn't hard to do—but the image was blurry. They'd have to sharpen the focus considerably to detect quasars, and even that would only be a step toward a fringe test, which itself was only a warm-up for the main event of observing Sagittarius A* and M87.

"I want to at least see 3C 273," Shep said to David. "If we can't see 3C 273 . . ."

Shep walked over to Gopal. "David is showing me the weather over the next week," Shep told him. "It looks horrible."

"Well," Gopal said, "these are the cards you're dealt."

"Thanks, Kenny Rogers," Shep said. "I want actionable intelligence, not aphorisms."

• • •

They ate dinner in the kitchen on the first floor of the observatory, and afterward they trudged up two flights of stairs to the industrial elevator that carried them to the control room. In the mountain atmosphere their feet were made of lead. The weather was bad and getting worse, but the Google document and EHT Wiki site now said that the other sites were observing. The LMT was not, because they still hadn't managed to focus the telescope.

The tau continued to rise. When it reached 0.6, Shep and Gopal decided there was no chance they'd be able to join tonight's fringe tests. Shep Skyped with Jason SooHoo, who was at CARMA, to tell him the news. Right about then, David came in the room. "It's hailing slightly, so I'm parking the telescope." A big part of a telescope operator's job is to know when to put the telescope away. Hail could damage the surface of the reflector, but there are other dangers. The LMT has a protocol to follow in case Pico de Orizaba erupts: evasive maneuvers to prevent volcanic ash and water from coating the dish in cement.

As soon as David parked the telescope, the tau shot skyward. With that, everyone walked down to the pitch-black parking lot, climbed into their government-issue trucks, and started the long drive down the dark switchbacks toward base camp.

MARCH 20, 2015

They returned to the summit before sundown. The outlook for the night, as registered on the EHT Wiki site, was grim. Only the SMA was ready to observe. Everywhere else, there were technical problems, bad weather, or both. The LMT's official status was "mediocre" weather, "should have occasional" opportunities to observe, and still "working on focus." The hope for the night was to do a three-millimeter test, followed by a fringe test with another telescope, and they needed to get it done before Gopal had to leave for the bus station before dawn.

The weather forecast looked bad for the entire observing window—through the twenty-ninth. Their best hope was that the forecast was wrong, or that the weather created a Mauna Kea–style inversion layer so it was clear on the summit but cloudy below, or that heavy rain and snow gave way to clear skies. Right now it was foggy and the temperature was just above freezing, and together those two conditions created a precarious situation. If the temperature dropped below freezing, the fog could freeze to the dish. The ice wouldn't hurt the dish, but it would make it impossible to observe until the ice melted, which probably wouldn't happen until the sun had been up for hours.

They spent the evening waiting on the weather to turn. A little after midnight, David, Gopal, and a few others climbed into the apex hole to check the dish for ice. When they shined a high-powered shop lamp out the hole, they saw fog flowing by like a river, inches away. "We're basically in a fog bank," Gopal said. It was impossible to do real science in a fog bank, and under normal circumstances, they probably wouldn't even keep trying to troubleshoot it. But Gopal had to leave, so the situation was dire. "We can try it," he said. "We just need to watch the temperature."

• • •

A little after 1 A.M., Gopal queued up Enya on the control-room speakers. Shep stared at him and raised his eyebrows.

"What's wrong with Enya?" Gopal said.

Shep smiled, shook his head, and turned to Jonathan and Gisela to talk about a technical issue.

Sail away, sail away, sail away.

"Hey, if you want to drive, you can put on something else," Gopal said.

"No, let's leave it on 'Sailing Away,'" Shep said. "It's awesome."

To this soundtrack, fortunes shifted. Gopal announced that they had just focused the telescope on Jupiter at the one-millimeter wavelength. The atmosphere was stabilizing.

"Now I'm getting pretty excited," Shep said.

Shep sat down at a table with his laptop, chatted with California and Hawaii, and told the room that the rest of the array was ready to go.

Now that the telescope was pointing and focusing properly, everyone was in high spirits—relieved, late-night, oxygen-deprived giddy. A student of Gopal's named Mike said, "I don't know if I want this to work, because I don't want Enya to be our lucky music. I don't think I can stand days of this."

As the beginning of the night's observing schedule approached, the astronomers quietly worked on their laptops. Lindy plugged his MacBook Air into an enormous blue cable coiled at his feet like a harpooning rope. Laura sat on an overstuffed black couch and typed commands to prepare the Mark 6 recorders and the digital back ends for the observation.

"Can I make a comment?" Shep said to Gopal. "How soon will we be able to fringe test? I don't want to let the perfect"—perfect focusing—"be the enemy of good."

"But they aren't collecting data yet," Gopal said, referring to the sites in California and Hawaii.

"But I can send an email and say let's do this *now*," Shep said. "Let's do it this moment before the tau goes crazy."

"Four out of seven sites are ready, including us," Laura said.

After a few minutes, Shep read the latest report from the other sites. "We're all ready to go."

He was about to send a note to the other sites instructing them to get ready when Gopal said, "We're way out of focus."

"Okay, what's the tau look like?" Shep asked.

"It's increasing," David said.

"We're still really out of focus," Gopal said.

"The schedule starts in forty-nine minutes," Laura said.

For the next few minutes, Gopal and David hunched over the control panels, trying to get the telescope in focus. After a while, Gopal slapped a triumphant drum riff on the table. "Focus is looking good!"

• • •

Forty-eight minutes later, the schedule, which involved oscillating between two quasars to see if this California-Arizona-Mexico telescope triangle could work together, was about to begin.

"Are we gonna be spinning disc pretty soon?" Shep asked.

"Yep, t minus four," Lindy said.

"It's game time, this is awesome!" Laura said.

"Are we ready to spin disc?" Shep asked, again.

"The disc will spin," Lindy said.

"The scheduler is running," David said.

"All right, thirty seconds," Lindy said.

"We're on 3C 279," David said. "Four, three, two, one, done."

"It's going," Laura said.

"I want to see lights," Shep said.

"Disc is filling, file size is increasing, and lights"—on the Mark 6s—"are blinking," Lindy said.

• • •

Two uneventful hours passed before Gopal delivered the bad news. "I'm starting to be worried about the optics of this thing," he said. "I think we could be way out of focus."

The usual move in a situation like this would be to drop out of the coordinated EHT schedule for a half hour or so and refocus the telescope on a planet. But Gopal had to leave in an hour. And he wasn't entirely convinced that the problem was with the focusing. "There are two things I'm worried about," he said. "One is that we may be way out of focus." That would indicate a problem with the LMT itself. "The other thing is that there might be something wrong in the optics in the dewar." That would be a malfunction in the cryogenically sealed innards of Gopal's makeshift Franken-receiver.

Gopal explained that inside the cryostat—the outer vessel of which is called a dewar—is the total power box. The total power box tells the telescope whether it's collecting the expected amount of light from the astronomical source its operators are trying to find. If the total power box is getting a noisy signal, it's hard to know whether the telescope is looking at the right thing, especially when that thing is something as faint and distant as a quasar.

"How do we determine which is the problem?" Shep asked.

"This is how we do it," Gopal said—meaning, we fire up the telescope and start looking at stuff. "We hadn't done it till now." Gopal made the case for dropping out of the observation schedule so they could troubleshoot the optics.

"Sounds like we should be focusing on the optics," Shep said.

"That's what I've been saying," Gopal said.

It was decided. "Let's drop out and go to Saturn," Shep said to David.

As Gopal packed up his stuff and prepared to leave, Shep paced between pulse-ox tests, testing his ability to elevate his blood oxygen count by careful breathing. "It's really unfortunate that you have to leave," Shep said to Gopal. "This is your system. What about Pete Schloerb? How is he with this stuff?"

"He's amazing with this stuff."

"Can we get him up here?" Shep asked. "We need him up here, like, now."

The focusing was off by a factor of two when Gopal put on his coat and backpack. "I will have Wi-Fi access on my Delta flight and I will pull the data and send you an email." Tomorrow night, he would join them remotely by midnight.

Shep was the director of the EHT, but this was Gopal's telescope, and that meant that in some sense, Gopal had been in charge. He understood the instrument better than anyone. Now Shep was in charge. What he had to go on was his experience with telescopes in general and a handful of exposures to this one. "Let's make a plan for the rest of the evening," he said to the room. "We've got forty minutes left."

"This is a good night," Gopal said as his parting words. Weather-wise, he meant. "Not a great night, but a decent night. My recommendation is we should try optimizing the focus as much as we can."

"Let's aim to be on source for the last scan of the night," Shep said.

"It's four fifty A.M., so we have a half hour," Laura said.

Gopal left. Shep paced the control room, glum and muttering. "I really wish Gopal didn't have to leave," he said aloud but mostly to himself. "He knows the telescope; he knows the receiver."

The last scan of the night was a brief observation of 3C 273, the first confirmed quasar, the pinprick light that inspired Donald Lynden-Bell to postulate the existence of giant black holes at the centers of galaxies. The tau was 0.3, the lowest it had been all night. Laura emailed the other stations to tell them to send ten-second samples of the last scan back to Haystack for correlation. The LMT and its partners in California and Arizona slewed in long-distance unison to the night's last target.

"So, that's the best focus I can do for now," David said to Shep, pointing at a monitor on the control panel.

"Do you think we're pointed?" Shep asked.

David sighed. "Within twenty arcseconds." Close enough for punk rock, but not for observing black holes. He paused, then added, "My concern is that I don't think there's anything really wrong with the telescope." If the problem was with the telescope—the LMT and its electronics or hardware or whatever else—that wouldn't be *good*, but they could probably repair it. They could call in help from Puebla or Mexico City to hunt down the problem and kill it. But if the receiver was a problem, well, that was a problem. The guy who built the receiver was probably just now boarding a bus in Ciudad Serdan bound for Mexico City.

"So you think there's something wrong with the receiver?" Shep asked.

"Well."

MARCH 21, 2015

They never expected to get an image of Sagittarius A* that year, because ALMA and the South Pole Telescope weren't involved. But adding the Large Millimeter Telescope to the standing array of proven Event Horizon Telescope sites was important. Once that was done, Shep could go to ALMA and say that if they got telescope time in the spring of 2016, they would be ready. Yet now, on day three of the observing run, they had yet to get the LMT to work. For reasons they hadn't figured out, they couldn't focus the telescope. Finding the source, or sources, of this failure was the priority for the night.

In the early evening, after a few hours of fitful sleep at base camp, the crew collected themselves for another attempt. Ciudad Serdan was most charming at this time of day. The soft evening light buffed away the village's rougher edges, and the townsfolk poured into the *zócalo* to eat, drink, sing, flirt. Pico de Orizaba and Sierra Negra usually loomed over the village, but this evening, they were hidden behind clouds.

Shep and David had a test planned, and it would require a little shopping. They stopped at the Bodega Aurrera supermarket for potato chips, bread, Styrofoam cups (for liquid nitrogen), and coffee mugs (for coffee). They drove from hardware store to hardware store looking for suitable wire—*alambre número doce, no cable*—that they could use to hang cups of liquid nitrogen between the telescope's mirrors and the receiver. By 7 P.M., when they started the drive up the mountain, the sun was down.

When they reached the summit, they headed straight for the receiver room. Shep twisted two wires together to form a long, rigid poker and stabbed a small rectangle of black foam onto the end. He poured liquid nitrogen from a large thermos into a Styrofoam cup, dipped the foam into the boiling cold liquid, then lifted the wafting nitrogen-soaked foam to eye level like he was inspecting a piece of fondue. "That's how you make a lollipop," he said.

He climbed into the mirror cabinet and picked up the Styrofoam cup of liquid nitrogen with his gloved left hand and the wire-and-foam lollipop with his bare right. He dipped the lollipop in the nitrogen and held it up to M4, the new mirror that was installed on their account.

Down on the floor level, a digital readout connected to the receiver lit up. Laura and Lindy read the measurements aloud as Shep explored the surface of the mirror with the lollipop.

"Eleven, twelve, fourteen, twelve, fifteen, fourteen," Lindy said.

"Getting warmer!" Laura said.

They were looking for M4's sweet spot, the point on the mirror that reflected the biggest slice of infrared light from the liquid-nitrogen lollipop into the receiver. The numbers that Laura and Lindy were shouting were temperature measurements; they wanted to find the coldest spot on the mirror.

"Let's systematize this," Shep said. "We're gonna go across here, on the long axis of the mirror. Write these down."

"First position," Shep said.

"Ten," Lindy said.

"Okay, next position," Shep said.

"Thirteen, fourteen," Lindy said.

"You started at the minimum," Laura said.

"Now I'm gonna go the other way," Shep said. "Position one."

After several minutes of this, Shep moved his cryogenic wand to a spot on the mirror that was well outside the center, and the temperature readout plummeted.

"Oooh, money, money, money!" Laura shouted.

Shep's eyes widened and his jaw dropped. "Are you serious?" he said. "Okay, shit. That's interesting. How can we be that far off?" For reasons that no one could understand, the mirror was horribly out of alignment. That would explain at least part of the focusing problem.

While Shep and the postdocs were performing their lollipop

routine, help arrived. David Hughes, the director of the LMT, was an imperious British astronomer who had been living in Mexico since he took the director's job in 2011. He was dressed in futuristic-looking activewear. Pete Schloerb, a professor at the University of Massachusetts Amherst, had been working on the LMT since the beginning. He happened to be in the area for the inauguration, yesterday morning, of a new cosmic-ray observatory at the base of this mountain. He wore a Carhartt jacket, Carhartt jeans, and a hunter-orange stocking cap, like a guy going deer hunting.

With help from David Hughes and Pete Schloerb, the astronomers broke out the lasers. Hughes placed small mirrors called optical flats on the mirrors, using lithium grease for an adhesive. Someone climbed into the vertex hole and powered up a red laser attached to the ceiling; the red laser pointed inward, tracing the path of incoming light. Shep and Pete mounted a green laser to the receiver; it would shine in the opposite direction of the red laser. Workers taped paper over the windows on the receiver room doors and turned out the lights.

If the mirrors were aligned properly, the red and green lasers would perfectly overlap. Instead, they were off by inches.

"Hold on," Shep said. "They did this"—meaning, a few weeks ago, Gopal and company performed the lollipop exercise—"and they did this," with the lasers. "How could they be off so much?"

"They said they lined it up perfectly," Lindy said. "Two lasers, both ways, aligned it, and it was spot-on. That was like three weeks ago."

"Something is not the same as it was, because the two lasers are not aligned now," Shep said. Lights out, lasers on, they slowly, gently nudged M5, the small mirror connected to the receiver. After about an hour, the lasers were aligned. Around midnight, Shep climbed back into the receiver cabinet for another session of lollipoppin'.

After a while, Schloerb delivered the verdict: "A lot better."

• • •

They took the elevator down to the control room and tried yet again to focus the telescope. Their instruments told them Jupiter was three times brighter than last time they tried, so the mirror adjustments helped—much more light was making it from the sky into the receiver. But they were still having trouble pointing the telescope on 3C 279, a much more evasive target.

By 3 A.M., everyone was quietly doing calculations by hand in notebooks or working on laptops. David Hughes queued up music by the indie singer Bon Iver, and as the astronomers patiently tried to detect 3C 279, a strangely apt chorus played over the control-room stereo: *And I could see for miles, miles, miles.*

It was becoming clear that the mirrors weren't the only problem. An unexpected source of noise inside the receiver was mucking up the data stream that made the telescope point. To get rid of the noise, they'd first have to find its source, and that could take anywhere from hours to days. After videoconferencing with the people at CARMA, Shep came up with a plan: build a filter out of spare parts and install it in the receiver to work around the noisy signal. David Sánchez was skeptical. Pete Schloerb was open to it. David Hughes started rummaging around for components. "We're going to need a big fucking capacitor," Shep told Hughes.

The astronomers took turns studying the receiver, trying to find its flaws. Eventually they found what they thought was the culprit: a loose piece of Mylar inside the receiver was picking up acoustic vibrations and turning them into noise. Shep was surprised, relieved, and bewildered. "So," he said, suppressing delirious, exasperated laughter, "what are we going to do about it?"

By 6 A.M., Shep had scraped up the parts to build his filter. Back in the control room, he sat at a table and soldered components together while David Sánchez and Pete Schloerb tried to focus the telescope on Saturn.

As tends to happen at sunrise after a solid night of grueling work at fifteen thousand feet above sea level, people started to hit a wall of fatigue. Shep wanted to keep going. Could they find other planets to focus on? Could they install the filter and see if that helps? David Sánchez could see that they were done. There was nothing to gain by working: the more they kept at it in their oxygen- and sleep-deprived states, the more likely they were to break something that wasn't already broken.

"We are too tired," David said.

"Okay, let's pack it up," Shep said.

The only one who showed no sign of exhaustion was Laura. She was just as chipper as she had been for the past three days. Because she was alert and needed to learn to drive the route between the summit and base camp, she took the wheel. She drove unnervingly fast at first, but by the time the scariest switchbacks were behind them, she had settled into it. David Sánchez sat shotgun and coached her on the right times to switch into low gear, to turn off four-wheel drive, to slow to a crawl to clear a *tope*. Shep sat in the backseat, and within minutes, he fell asleep, head hanging forward, swinging with the jolts of the road, chin touching chest, as if shot by a sniper.

21

In the days after the long night of the liquid-nitrogen lollipop, they isolated two separate problems. The mirrors were misaligned because M4's mount was broken. That was easy enough to fix. The reason they couldn't point the telescope was, sure enough, a glitch inside the receiver's helium-cooled dewar. To this problem, there was no clear solution—at least not until Gopal came back. No way was Shep or Laura or Lindy going to open the helium-cooled cryostat of Gopal's hand-built receiver. They tried writing software to overcome the noise, but it wasn't quite enough, so they came up with a tedious workaround. First they'd use the three-millimeter receiver to point the telescope, then transfer the pointing data to the one-millimeter receiver, and go from there. This meant that they had to manually remove and then reinsert M4 every hour. But it was good enough to get them through the week.

The skies cleared on the last night of the run—and then conditions were good enough that they decided to add another, bonus night—but in general, the weather remained awful. All the data from the run was compromised in one way or another. After about a month, they got fringes between the LMT and the SMA, and they gave themselves credit for fielding entirely new equipment at every site in the array. But no scientific discoveries were likely to be waiting in the two hundred terabytes of data they shipped back from Mexico.

• • •

Two weeks after the end of the spring campaign, Shep and the family drove to New York so he could give a lecture at the American Museum of Natural History. It was a warm, clear early spring evening on the Upper West Side. Shep spoke to a sold-out crowd inside Hayden Planetarium, his slides projected onto the domed ceiling.

Standing at a podium in the center of the darkened room, he delivered a polished version of his usual talk, enlivened for the public. "It's the holy of holies for an astronomer to be invited to Hayden Planetarium," he began. "Like a priest being called to the Vatican." Black holes, he explained, are where general relativity and quantum mechanics "shake hands." If we look at them with "radio goggles," we see jets of material shooting out of them, propelled by an explosive force equivalent to "ten billion billion H-bombs going off every second." "Bruce Willis was unavailable to pilot them there on a ship built by the administration of Morgan Freeman," he said. "The only processes known for creating this energy are spinning supermassive black holes."

He described how matter collapsed to unimaginable densities—neutron stars were "the size of New York"—and how supermassive black holes were probably formed by the merger of galaxies. He showed a video simulation of such a merger, and warned, "This is pretty graphic—it shows the full courtship and consummation, so if there are any children here, you might want to cover their eyes."

In a pretty dramatic contrast with those recent nights on the mountain in Mexico, everything at the planetarium was breaking his way. He showed simulations of the shadow of Sagittarius A*, and he played an animation that Laura Vertatschitsch made: a globe, dotted with the telescopes that make up the EHT, rotates, and as it does, baselines between the telescopes appear, "as though we have these spiders working for us, spinning this gossamer" between them. He described some of the fieldwork, and showed a slide from the previous year's maser installation at the LMT. "If you haven't had the opportunity to watch a three hundred thousand dollar atomic clock hoisted by a slender cable up a helical staircase," he told the audience, "you haven't lived."

He explained the plans for the next two to three years. Grow the number of stations from three to eight. Increase the amount of data recorded by sixteen times, and expand the collecting area of the EHT tenfold. Get a picture of the shadow.

When he took questions, the first to raise his hand was an enthusiastic college-age guy. "I just saw your BBC documentary," he said to Shep. "You're amazing. What do you think happens at the singularity?"

Afterward, Shep, Elissa, and the kids lingered outside the museum and then walked down the street to a restaurant on Columbus Avenue. Shep had an expression of happy, exhausted relief: for the first time in a while, the night was a win.

• • •

The people who ran ALMA were a less adoring audience.

Back in March, on the first day of the spring observing run, Shep called Tony Beasley, the director of the National Radio Astronomy Observatory, from the LMT base camp in Ciudad Serdan and learned that ALMA had imposed some new conditions. ALMA now expected the Event Horizon Telescope to offer its services to any astronomer who wanted to write a proposal. The EHT would have to open itself up to the entire astronomical community. In theory, someone could come out of nowhere, write a Sagittarius A* proposal that ALMA liked better than the EHT's, and then Shep and company would be forced to run *that* person's proposal. The logic went like this: ALMA was built by the international community with a strong policy of "open access," meaning that any capability that ALMA has—including the ability to expand into an Earth-size virtual telescope—must be available to the global astronomy community. Ergo, the Event Horizon Telescope must be available to the global astronomical community. From the whitewashed stucco base camp in Ciudad Serdan, to an ambient background of baying dogs and car-top loudspeakers, Shep listened as Beasley explained the new terms and thought: *Every syllable of this is insane.*

Shep hung up the phone and thought: *I need a break*. But there was no time for a break, even after the brutal fortnight in Mexico. The weeks after the campaign were spent struggling to wring data from the disk drives. Any other year, late spring and early summer were a time for postobservation detox, for reflection and planning, for reconnecting with spouses and kids and pets. This year, all the data they sent home was compromised in one way or another. Not just the data from Mexico. CARMA had problems getting data off disks. SMA was recording in a different format from the rest of the telescopes, so they'd have to convert it before they could do anything with it. They had fielded *so* much new equipment, *so* much new software, that they had no fixed points of reference.

Every time Shep checked on the ALMA situation, it seemed to get worse. In May, at a meeting of the National Radio Astronomy Observatory board in Green Bank, West Virginia, he met Tony Beasley in the observatory lounge and tried to explain why ALMA's demands were unreasonable. Beasley wouldn't have it. *You're pissing people off with your intransigence,* he told Shep. *You guys are running a seat-of-the-pants operation,* he said, *and ALMA will never let you near their telescope as long as you are.* His message: face reality. You must accept ALMA's demands. You'll be lucky to get on ALMA in 2017 even if you do.

Shep felt as if he were competing in a sadistic game show in which each time he completed an obstacle course, he'd have to do the course again, only this time great white sharks would be involved.

Mike Hecht kept asking Shep: what do you *want*? Shep said he wanted a situation that respects the contributions of the people who have been doing this for years. He felt that Heino wanted to buy his way into the project and create a false history of the black-hole shadow in which he was the prime mover and main discoverer.

Shep came to realize that his real argument didn't persuade anyone. His real argument was this: it's not fair. It's not fair for him to build this thing and then be forced to give it away. He's spent decades—his entire career—and raised millions of dollars to do this one experiment. Now he's expected to give it away—to Heino, to

ALMA, to any random astronomer who wants to look at something with an Earth-size virtual telescope?

It was natural to wonder whether they needed ALMA. Jonathan Weintroub, in exasperated moments, was known to float the possibility of cutting ALMA out and using the telescopes they had. Michael Johnson and Andrew Chael had been doing computer simulations to see whether they could extract an image from data they had already collected. The answer was: maybe. If they had the Atacama Pathfinder Experiment, APEX, they'd have baselines to northern Chile, which would give them the same geographic coverage they'd have with ALMA. Of course, ALMA was the most powerful radio telescope on Earth, so they'd have to make up for the loss of sensitivity somehow. They might be able to do so electronically, with superfast recorders and high-capacity hard drives—they might be able, in Shep's terms, to "wideband the fuck out of it." It would be such a poke in the eye to get the picture of Sagittarius A* without ALMA. But Shep never sounded more than quarter-hearted when he talked about this revenge fantasy.

• • •

Back in 2012, Shep won a Guggenheim Foundation grant that he planned to use to move the family to Chile for a year. They'd live in Santiago, send the kids to school there, and Shep would take an office at ALMA headquarters, show up five days a week, and ingratiate himself with those in charge. The Chile thing never happened, because it's not easy to relocate a two-career family with preteen kids to a foreign country for a year, especially while trying to direct the construction of an Earth-size virtual telescope, so Shep never collected the Guggenheim money. When 2015 came around, the money was still available, and the people at the foundation were surprisingly relaxed when Shep told them that instead of spending a year in Chile, he'd like to take the family to Hawaii for six weeks. They rented a house outside Hilo, and Shep reported for work each day at the Submillimeter Array headquarters in the University of Hawaii research park, a collection of sea-level offices that support the big telescopes on Mauna Kea.

They worked. Shep and Geoff Bower, who was living in Hilo and working for the Academia Sinica Institute of Astronomy and Astrophysics, or ASIAA, the Taiwan-based research institution that managed the James Clerk Maxwell Telescope, tried to do the highest angular resolution ever made from the surface of Earth—a 0.87 mm observation between ALMA, APEX, and the SMA, on Mauna Kea. It failed, because of—naturally—a software error.

They also played. Shep and Elissa and the kids took a seven-day bike tour of the Big Island's west coast, pedaling north from Kona, circling the base of the mountain where their dad's career had become the seething, swirling storm it was today.

Distance put the organization-building process on hold. Holding a teleconference that includes people in Asia, Europe, and the United States is feasible as long as someone gets up at 6 A.M. or stays up until 11 P.M. But Hawaii might as well be in its own parallel universe. People there have to wake up at 3 A.M. to join a worldwide conference call.

Near the end of their stay, Shep flew to Honolulu for a meeting of the General Assembly of the International Astronomical Union. Anton Zensus and Tony Beasley would be there, and Shep wanted to make nice. One of the big sessions at the meeting was called "Celebrating Radio Astronomy's Golden Years," and it was about . . . the 1960s. Shep thought it was a bad title, because it said, *You guys missed out. Anyone who isn't this old by default did not participate in the golden age.* And yet here he and Anton Zensus and Tony Beasley were in Honolulu, plotting one of radio astronomy's greatest coups.

There was never a moment he could isolate, but the break, the distance, the sunlight, and the quiet—they worked. Everyone noticed the difference when Shep returned from Hawaii. He was ready to make a deal.

In September, Shep received a letter saying that ALMA's final deadline for getting submillimeter VLBI approved as an official observing mode for Cycle Four, the block of telescope time than ran from late 2016 through the spring of 2017, was at the ALMA board meeting in November. If they wanted to light up the Event Horizon Telescope for the first time in 2017, they had to get on Cycle Four.

Shep had discarded any fantasies about getting director's discretionary time or other special treatment. He'd accepted that they'd have to get in line like everyone else. What he and others, especially telescope directors who didn't like ALMA telling them what to do, had spent the past several months processing was the demand that the EHT become an open-access facility. Now they had a deadline. By November, Team Shep and Team Heino would have to agree on a plan for merging the Event Horizon Telescope and BlackHoleCam. They would have to accept ALMA's terms. And they would have to present a united front to the ALMA board. If they didn't, they could forget about taking a picture of Sagittarius A* before 2018.

Shep still had reservations. As soon as you agree to a structure in which the director of the organization reports to a board, nothing is guaranteed. As for ALMA's demands, there was always the possibility that some random astronomer could swoop in and beat the EHT astronomers for observing time on their own array.

He'd been war-gaming every imaginable scenario in which he might lose control. *Maybe,* he began to think, *the reason people have been pushing back is not that they're trying to screw me over, but that these scenarios are so incredibly remote that it's not worth holding up the entire project until we can figure out how to preempt them. Maybe it's not in anyone's interest to have this go on any longer.*

People know what they want at the end of the day: credit. To have their surname mentioned in association with the first picture of a black hole for as long as civilization lasted. There was no single lever to pull or button to push that would achieve that. Everyone had an incentive, as they designed this organization, to build in as many levers and buttons as possible. That, plus cultural differences and decades of contempt-breeding familiarity, created an atmosphere of distrust.

The risks, he thought, were tremendous. But more and more, he just wanted to get this done. The EHT now had an office crowded with young researchers, graduate students and postdocs who were counting on this telescope firing up before their appointments were finished. They needed the EHT to start taking real data so they could write papers that would launch their careers. If this dragged on for another year or two or three, those people would leave the EHT empty handed.

A few people still argued that they should pursue the rogue scenario—go for the image without ALMA. But Shep was newly interested in the art of the possible. *Widebanding the fuck out of it would make for a great story,* he thought, *but planning for a great story was not planning for success.*

PART FOUR

THE EARTH-SIZE TELESCOPE

22

JANUARY 2016

Just after the New Year, Remo Tilanus sent an email to the collaboration. They had fourteen months before the big observation, and they were already behind. Because they had finally agreed on organizational principles in the fall, ALMA had approved VLBI as an "advertised capability" for Cycle Four. Unless something catastrophic happened, they'd add ALMA to the Event Horizon Telescope in 2017. But they'd spent so much time on politics that they'd have to hurry. Fourteen months would pass quickly, and there was much to do. They still had to submit a formal proposal to ALMA asking for observing time, for one thing.

Last year, in the rush to do one last observation with CARMA, they had installed new VLBI equipment at every site, and despite the chaos of the spring 2015 run, when they finally processed the data, they found that everything worked as it was supposed to. But that didn't mean it would work next year. The telescopes themselves were making all sorts of upgrades in the coming months, so they weren't working with a stable platform. They had to observe again this spring, in a practice run for 2017, and then they'd have to do several more test runs before the big event—at minimum, one in the fall and then another one again in the winter.

Remo's priority as someone who was "in charge"—he was still

the interim project manager, so he was responsible for looking after things, but didn't have any real authority—was to solidify the core of the array. The way he saw it, as of January 2016, there weren't many telescopes they absolutely could count on. The Arizona station, the SMT, was one. He wanted to add the Hawaiian and Mexican stations to that list. They'd use the solid core for a series of tests with ALMA, and thereby avoid embarrassing themselves. They'd build out from there, bringing more telescopes into the pool of well-tested, well-seasoned instruments. APEX. Pico Veleta. The South Pole Telescope, which Dan Marrone and his team would equip on a return trip in December. They would equip and test and troubleshoot these telescopes in an orderly sequence, on an orderly schedule, and they would make sure they looked like professional astronomers who don't break everything they touch, because ALMA was watching.

• • •

Shep was settling into his office in the Harvard-Smithsonian Center for Astrophysics building on Concord Avenue. The once-barren room now wore the beginnings of decor. On the wall adjacent to his desk, a yellow expressionist floral print that he bought at a local artist's open gallery for eighty dollars. On the exterior of the entry door, a printed photo of a 1995 *Weekly World News* tabloid story about how scientists had discovered that Earth was orbiting the sun at 66,666 miles per hour—the speed of Satan! Shep's favorite part of the article was an inset illustration with the caption, "Artist's rendering of Satan in the flames of hell." Shep had a thing for the *Weekly World News*. His biological father once wrote an article about the paper, about how they recruit good writers and pay them well, because once they crawl in the lowest realm of tabloids, there is no going back to legitimate journalism.

Another new object in his office was a copy of John Templeton's book *The God Who Would Be Known*. Templeton was an early Warren Buffett figure, a stock-picker who made billions in the decades between World War II and his retirement in the 1990s. When it came time to give away his fortune, he founded the John Templeton Foundation, which aimed to "advance human well-being by supporting research on the Big Questions," and, in the process, to reconcile religion and science. Templeton was a complicated guy. He was a ruthless capitalist who in the late 1960s renounced his American citizenship and moved to the Bahamas to avoid taxes—and then argued that this was for the greater good, because the money he would have paid in taxes he could now use for philanthropy. He was a devout Presbyterian and a devoted patron of science—particularly, science that probed the mysteries of the mind and the fundamental nature of reality. He didn't take the Bible literally. He thought that humans were profoundly ignorant of the true nature of the universe, and that both science and religion brought us closer to the truth. To facilitate that progress, he set up a foundation to support scientists exploring questions such as: "Does nature offer evidence of purpose? What is the nature of spacetime? Do we inhabit a multiverse? How did life originate?"

And that is why Templeton's book was on Shep's office shelf. Shep had been talking to the foundation about funding a Black Hole Initiative at Harvard, an interdisciplinary center for astronomers, physicists, mathematicians, and philosophers. It would be the first center of its kind in the world. He had tried to launch the center with a grant from the National Science Foundation. Last year, he teamed up with Avi Loeb, Ramesh Narayan, and a small team of other luminaries: Andrew (Andy) Strominger, theoretical physicist; Shing-Tung Yau, a mathematician who made discoveries that were foundational to string theory; Peter Galison, a philosopher and historian of science. The night before the deadline, Shep and Michael Johnson pulled an all-nighter at the Center for Astrophysics and filed the proposal at 1 P.M. Shep then walked downstairs, got in his car, started to drive home, and, within minutes, fell asleep at a stoplight. He woke up to police lights in the rearview mirror. Cars were speeding around him. The officer asked him, *Were you on the phone or something?*

No, um, I was asleep.

Exasperation. *If you're that tired, do you think you should be driving?*

Shep couldn't find his registration in the rat's nest of his glove box, so the cop looked it up. His inspection certificate was three years expired, so he got a moving violation for it. That ticket and others, plus the move to the Center for Astrophysics, turned Shep into a bike commuter.

Shep always said that he's "basically" an atheist. He and the family went to temple and lit candles on Friday nights. He'd describe their practice as Judaism Lite. But he liked the ritual, the tradition. And for all his discomfort with religion, Shep approached matters of the spiritual with respect bordering on awe. On the family trip to Israel a year and a half ago, he was moved by the Church of the Holy Sepulchre, which, per tradition, holds the site of Jesus's crucifixion and burial. Shep watched the pilgrims approach the aedicule, the site of Jesus's tomb, with reverence. He enjoyed talking about God and the cosmos. He and an old rabbi used to commiserate, talk shop. Astronomy and religion, they agreed, were both ways of looking into the past. A telescope was an instrument for witnessing the ancient, for receiving messages transmitted by matter undergoing change thousands or millions or billions of years ago. Ritual was a means of communing with lost generations, of connecting to the past by repeating, with focus and intentionality, the actions of your ancestors.

So Shep was okay with the idea of taking money from the Templeton Foundation. Besides, he had to be practical. In his talks and proposals and pitches to funders and science-policy types, he tried to position the EHT as something that could, and should, keep going long after the first big observation. If they saw the shadow of Sagittarius A* in the data they hoped to collect next year, great—but they could take an even better picture in 2018, when they had doubled, yet again, the bandwidth of their telescopes. And they could keep going. They could build new telescopes, in Greenland, in Africa. They could push the bandwidth as high as 256 gigabits per second, ten thousand times the average broadband Internet speed in the United States. At that level, they wouldn't need the biggest, highest, most advanced dishes in the world—they could use smaller dishes and make up for the loss of collecting area with a fire hose of data. Even so, telescopes become obsolete. Experiments conclude. Shep was beginning to accept that this thing would run its course, and that it was time to prepare for life afterward.

23

In April 2016, for the first time since grad school, when he synchronized his life's schedule to the annual Sagittarius A* observing cycle, Shep decided to stay home during the spring observing campaign. He had been planning to go to Mexico for the run, but Elissa had to make a work trip to Washington, D.C., and they already had plenty of people going to the LMT—Lindy, Gopal, Jason SooHoo. They could handle it. It was really a practice run for 2017. Plus, there was plenty to do in Cambridge.

Even after last fall's detente, it took months to come up with a collaboration agreement everyone could agree on. The distrustful part of Shep's psyche still had moments of resurgent strength. Every time they settled one issue, another would emerge. Should theorists be on the board? Should they be a "stakeholder" in the same way as someone who brings a telescope into the array? What does a telescope have to bring? Do they have to bring guaranteed time? Shep wanted to game out every conceivable scenario, prepare for every contingency, construct an airtight charter that protected his ownership stake in the EHT no matter what. He'd been reading about how organizations work; he'd realized what was missing here was trust. Everyone knew that if they didn't come together, the project would fail. A few dozen people from different countries and different scientific cultures with different institutional pressures and policies would have to find common ground, or else the black hole at the center of the galaxy would go unseen. That clarity didn't make things any easier. The stakes were too high. At one point, Shep was talking to professional facilitators for advice, thinking he might need to bring in an arbitrator. Maybe that's what they needed. Astronomers were notoriously indecisive. You can always recognize astronomers at a conference, Shep liked to say, because around dinnertime, they're not inside a restaurant eating, they're outside dithering over where to go. Asking these people to handle a delicate international negotiation was like turning a human torch loose at a gas station.

And then there was the question of how, once the collaboration came together, they would deal with ALMA. Shep was still worried that, under ALMA's open-access rules, some outsider could swoop in and steal their observation away from them. Every time he brought it up, Shep heard the same thing: if someone else writes a Sagittarius A* proposal and it's ranked higher than the EHT's, the EHT is out of luck. If Shep and company couldn't write the best Sagittarius A* proposal in the world, they didn't deserve to use their own creation—which, after all, relied on billions of dollars of telescopes around the world that they didn't build.

These issues and others were all coming to a head in April. They did, ultimately, come up with a collaboration agreement that they put to a vote. The last day to vote was April 1. They still had to apply for time on ALMA, and that proposal was due in late April. There was another big April event Shep had to prepare for. On April 18, Stephen Hawking was coming to town for the inauguration of the Black Hole Initiative at Harvard.

• • •

On a sunny afternoon in April, hundreds of students, professors, and members of the general public queued up under the sixty-foot ceilings of Harvard's Memorial Hall to watch Stephen Hawking give a talk called "Quantum Black Holes."

In the north end of the memorial transept, Shep was schmoozing. He was wearing the astronomer equivalent of black tie—a tan blazer and chinos. Elissa waited while Shep shook the necessary hands. She was in a good mood, not only because she, like most people in the hall, was receiving the vibrations of the crowd. She'd just been promoted to associate professor.

A half hour or so before the talk, they were ushered into Sanders Theater, a thousand-seat chamber of high seriousness. "This is kind of a circus!" Shep said to Tony Beasley. Stephen Hawking rolled onto the stage, and the room fell into a deep silence. His medical equipment laid down a cadence that reverberated throughout the acoustically perfect room: *Whirr, whine, beep. Whirr, whine, beep.*

The famous computerized voice spoke. "Can you hear me?"

"I am pleased to be here at the inauguration of the Black Hole Initiative," he began.

The Black Hole Initiative was not, technically speaking, ready to be inaugurated. The Templeton Foundation hadn't yet approved the grant that would fund it. The whole reason the inauguration was happening now, and not, say, when the Black Hole Initiative's existence was guaranteed, was that Stephen Hawking was already scheduled to be in the United States to launch a project called Breakthrough Starshot, an effort, funded by the Russian billionaire and philanthropist Yuri Milner, to send a tiny laser-powered space probe to Alpha Centauri. Getting Stephen Hawking across the Atlantic required an ambulance jet. Avi Loeb, the Black Hole Initiative's director, wanted to take advantage of the North American landing of this rare bird for the initiative's inauguration. So he planned the event and then asked the Templeton Foundation to please hurry and make up their minds. And now Stephen Hawking was onstage in front of a thousand people congratulating the Black Hole Initiative on their success.

But what mattered in the moment was that the great man himself was here, explaining the mysteries he had spent his career pondering—particularly the deep, fertile connection among black holes, thermodynamics, and information theory. The greatest among them was the black-hole information paradox, which, as he explained, "strikes at the heart of scientific determinism." He had been searching for a solution for forty years. "Finally," he said, "I found what I think is the answer."

Working with two other theoretical physicists, he was beginning to understand how a mechanism called supertranslation might encode information on the horizon of a black hole. "Watch this space," he said. Yet he already felt confident that black holes were not "the eternal prisons we once thought." "If you fall into a black hole," he said, "don't give up. There is a way out."

After the talk, the audience filed out while the people who had been sitting in reserved seats lingered near the stage. Shep, Avi Loeb, Ramesh Narayan, Andy Strominger, Peter Galison, and Shing-Tung Yau—the six senior faculty of the Black Hole Initiative—climbed onto the stage and posed for pictures with Hawking, an "honorary affiliate" of the center. Then the crowd began the walk across campus to the Harvard Art Museums, which Avi had commandeered for a celebratory dinner.

Round tables were set in the museum's ground-floor courtyard, under a glass Renzo Piano ceiling. Shep and Elissa were seated next to Jonathan Weintroub and his wife, Robbie Singal. "Shep and I need therapy, why are we seated next to each other?" Jonathan said when he saw the place cards. Right about then, Shep walked up. Jonathan was wearing a vaguely Hawaiian button-down shirt. "Nice style, Jono," Shep said. "You've got your shirt unbuttoned to your sternum."

• • •

The not-quite-final Templeton Foundation grant that these events were designed to celebrate consisted of $7,204,252 to use over three years. The hope was to find a wealthy backer to start an endowment that would keep the center going in perpetuity. Already, though, the initiative had elevated Shep to a new rung on the Harvard Square hierarchy. He was no longer a mere instrument builder. Because Harvard agreed to renovate five thousand square feet of prime real estate to house the Black Hole Initiative, Shep now shared an office with some of the biggest and most productive names in the search for the ultimate laws of nature.

In his talk at Memorial Hall, when Stephen Hawking said he

finally had the information paradox cracked, he was referring to his work with two people: Malcolm Perry of Cambridge University, and Shep's new colleague Andy Strominger, a theorist famous for his transformative work in mathematical physics. In the 1990s, he discovered a set of six-dimensional mathematical objects, which he named Calabi-Yau spaces—the "Yau" in that name being a nod to another member of the Black Hole Initiative, the mathematician Shing-Tung Yau—that could explain how string theory could describe a four-dimensional universe filled with the type of matter we see in ours. The discovery turned string theory from a dead-end basement hobby into the great hope of theoretical physics. For the next two decades, Strominger fished the fecund seas of string theory and its successor, M-theory, in which the "M" stands for "mystery," "matrix," or "membrane," depending on whom you ask. In 1996, he and Cumrun Vafa showed that string theory could account for the microscopic constituents of black-hole thermodynamics: the temperature of a black hole could be described by strings jittering away in hidden extra dimensions. By the time he joined Shep and Avi Loeb and three others to form the Black Hole Initiative, Strominger had become the elder hipster of the Harvard physics community. He wore chunky black glasses and black T-shirts and jeans. He spoke softly and deliberately, as if translating on the fly from a foreign language, that language being algebraic geometry. From a high-ceilinged, wood-paneled corner office on the top floor of Harvard's Jefferson Laboratory, he directed the Center for the Fundamental Laws of Nature, where eight professors and some forty postdocs and graduate students probed the deepest recesses of reality.

The Strominger-Hawking-Perry argument was that the no-hair theorem, the generally accepted idea that a black hole can be completely described by its mass, angular momentum, and electric charge, is flawed. In fact, the researchers argued, black holes have "soft hair," and through the mechanism that Hawking mentioned in his talk at Memorial Hall—supertranslation—this soft hair records information about particles that have fallen in.

Soft hair didn't mean much for Shep. But another of Strominger's ideas did. Strominger was convinced that evidence of strange new phenomena was raining down on us in the form of radiation from the edges of black holes. He was particularly interested in a mathematical property, known as conformal symmetry, that appears in all sorts of systems, including water and magnets, close to phase transitions. In a special class of black holes—those spinning at nearly the speed of light—spacetime undergoes a change that is much like a phase transition.

In the early 2000s, Strominger started on a program to use this clue—conformal symmetry—as a lead for pursuing a quantum theory of gravity. Just as the macroscopic properties of a vessel of water—its temperature, viscosity, and so on—are our blurry perception of mass behavior of trillions of atoms, so are the macroscale properties of spacetime. Here we return to the Big Question: what are those atoms? We don't know. But if you want to find out, and you identify a situation in which spacetime itself seems to behave *exactly* like water and magnets and other systems at their critical points—points that are all governed by the same basic set of mathematical rules—then you know you're on the right track.

You can pursue this lead in two directions. One is to hunt for the atoms of space—to join in the long-running search for a quantum theory of gravity. Another direction is to see what these strange properties of spacetime do to the stuff nearby, and how those effects might show up in astronomical signals—for example, in the light collected by the Event Horizon Telescope.

Astronomers calculate the spin of a black hole by studying x-rays emitted very close to its horizon. When a black hole spins at near-maximal speed, light emitted from close to the event horizon gets stretched toward the infrared end of the spectrum—redshifted—almost to infinity. Scientists can measure redshift with exquisite accuracy. Over the past decade, they have applied this measurement to x-rays from a couple dozen astrophysical black holes, and they have found that most of those are spinning rapidly, many at nearly the speed of light. The black hole GRS 1915+105 is spinning at 98 percent of the speed of light. MC 6-30-15 is nearly maxed out at 99 percent of the speed of light. M87, one of the Event Horizon Telescope's two main targets, wasn't spinning that fast, but fast enough to be a target. In these and other black holes, there seems to be a relationship between spin and jets, those cosmic fire hoses. The faster the spin, the thinner the jet. By the messy standards of the real world, that is freakishly perfect. Strominger and others suspected that these jets were critical phenomena written across the sky.

On a large chalkboard in his office, Strominger has written a list taller than himself of specific, tractable problems to be solved—relatively doable chunks of work, each enough to occupy a talented student for anywhere from a few months to a few years. One of those students, a doctoral candidate named Alex Lupsasca, along with the University of Arizona physicist Sam Gralla, took on the task of determining what the Event Horizon Telescope was likely to see when it looked at a black hole spinning at these near-maximal speeds. Their predictions looked completely different from the simulations the EHT's theorists had been producing.

Lupsasca set out to find what an observer looking at a near-maximally spinning black hole would see if some bright object—a star or a hotspot—were orbiting it. For reference, he first constructed a model that ignored the spacetime-warping effects of general relativity. In it, the observer sees exactly what our intuition, shaped by our understanding of moons orbiting planets and planets orbiting the sun, would expect: a black circle—the shadow—with a bright dot transiting from left to right along its equator. When Lupsasca switched on the effects of general relativity, however, everything went sideways. Literally sideways. Instead of transiting peacefully along the shadow's equator, images of the star or hotspot shoot upward along a vertical line off to the side of the black hole shadow.

The reason for this odd effect is that the orbiting star or hotspot *isn't* transiting peacefully along the equator of a dark circle in flat space: it's spinning around the inside of the throat of the black hole. In that region, everything is circling the throat in the same direction at the same speed. Periodically, an image of the star or hotspot gets flung out of the throat like a stone from a sling.

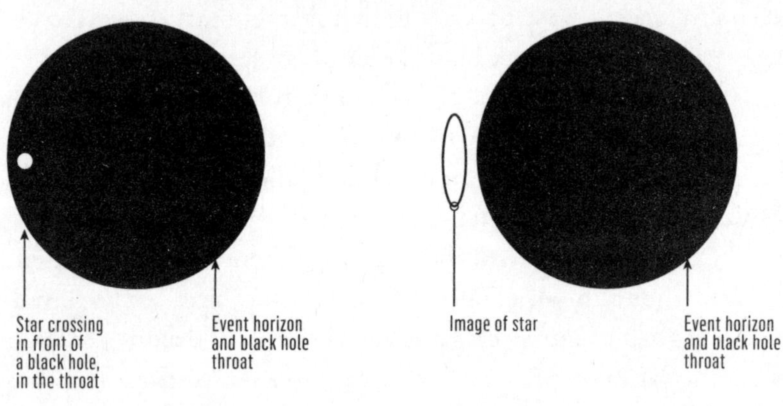

If the Event Horizon Telescope were to observe this bizarre spectacle, it would provide evidence that near the horizon of rapidly spinning black holes, conformal symmetry holds. That, in turn, would lend credence to the existence of the "holographic plate"—the two-dimensional region described by a two-dimensional conformal field theory—that theorists since the 1990s have argued must store information about the contents of a black hole.

24

Shep now had three offices, which was one too many. That summer, he decided to leave his half-time role at Haystack for the full-time job at the Center for Astrophysics.

Technically, you're supposed to work at the Massachusetts Institute of Technology for twenty-five years to get a commemorative office chair when you leave. Shep had been at Haystack for twenty-four years, twenty-one of those as a legitimate full-time employee. They gave him a chair anyway.

They threw his going-away party in the main Haystack conference room, the place where Shep had at least once said the pressure of his job was going to compress him into a diamond. They had cake, coffee, gifts. Alan Rogers, Mike Titus, Colin Lonsdale—the people who taught him, worked alongside him, and sometimes fought him—gave speeches, and in his remarks, Shep thanked them, and others, in turn.

He left feeling good about things. He was also terrified. He thought of scientists as organisms who live in a terrarium. In grad school, Haystack was the first place he found that had the right combination of atmosphere and nutrition. Could he thrive in another terrarium? He worried about that. He'd been working half-time at the Center for Astrophysics now for three and a half years, but he had yet to get over his imposter syndrome. He worried that people thought he was only able to do what he'd done because he happened to be at Haystack. He now had a chance to prove otherwise. But he'd have to learn to operate in a new ecosystem.

CAMBRIDGE, MASSACHUSETTS
NOVEMBER 2016

As the months passed, Shep's worst fears failed to materialize. In July 2016, the jurors who allocate time on ALMA met in Vienna and gave the EHT most of what it had asked for. Shep was annoyed that some of their minor proposals were turned down, but the important ones—particularly the main Sagittarius A* track—were approved. The dreaded open-access ALMA proposals—the ones Shep worried would steal his life's work—turned out to be harmless.

There was always some new political complication to keep Shep agitated. But anyone who had attended the Waterloo meeting two years ago could sense a shift. The collaboration was growing a new identity, one untethered to the dramas of the past. The hip-looking students and postdocs milling around the lobby bar in the evenings had little knowledge of and no use for the conflicts between Shep and Heino. They were just excited to be here.

On the second-to-last evening of the conference, Katie Bouman and Andrew Chael taught the assembled scientists how to make images from raw data. They took over the hotel-lobby dining room and pushed tables together to form a huge common work area. A couple dozen young astronomers, many from the European camp, showed up for the tutorial. Shep sat at the middle of the table with his laptop in front of him. It was cocktail hour, and the room was loud and festive.

To test Katie's CHIRP algorithm against other methods, Katie, Andrew, and a few others had created an open-source EHT imaging challenge. Anyone could go online, download real and synthetic EHT data sets, run them through CHIRP or their own imaging algorithms, and compare the results. In the cocktail-hour workshop, Katie and Andrew walked people through the imaging challenge. Katie and Andrew circled the room and talked over the background chatter.

About halfway through the workshop, Heino stopped by. He and Shep were both smiling, relaxed. Heino was happy to see how many of his students and postdocs had shown up for the tutorial. Shep was happy that his student was teaching Heino's students. For a moment, everyone was having . . . fun.

LARGE MILLIMETER TELESCOPE JANUARY 2017

Throughout the fall and winter, the telescope blinked on, baseline by baseline, each successful test adding another filament to its light-catching web. In November, APEX and SMT joined hands: one Chile-to-Arizona baseline down. The following month, Dan Marrone, Junhan Kim, and Andre Young traveled to Antarctica to prepare the South Pole Telescope. A few weeks later, crews flew to Mexico and Chile to conduct a test that would unite a long-sought triumvirate of telescopes—the SPT, ALMA, and the LMT.

Shep had wanted to go to the pole. He would have turned fifty years old there. It would have made for a pleasingly symmetric mid-life completion of his arc. The pole was crowded, though. Lots of projects were nearing completion, and there were more people with valid reasons to visit the pole than there were beds. Plus, Shep had too much to do in Cambridge to spend two months on the ice. He had an experiment to direct, a new academic center at Harvard to help build, and a wife and two teenagers. Instead, in late January, on his fiftieth birthday, Shep once again found himself on Sierra Negra, trudging up that godforsaken helical staircase in the thin, sterile air, boarding the gleaming freight elevator, and ascending to the control room of the LMT.

It was early on the first evening of the test run. The back-end room was quiet except for the tiny-turbine whine of the cooling fans in the Mark 6 recorders. Shep and Lindy were calmly examining cables and typing commands into laptops. Shep was wearing his usual light black jacket, which was a good sign, because yesterday he was sheltering inside an enormous down parka. He'd been sick with a fever and wet cough for more than a week. A doctor probably would have advised a person in such condition against flying to Mexico and working overnights at fifteen thousand feet above sea level. Just in time for the big test, though, he was on the mend.

Next door, Gopal and David Sánchez were keying in their own commands. Mendelssohn was playing on the control-room speakers. The screen in front of David showed clouds rolling over the summit, which didn't worry anyone, because it was still early; later, the clouds would descend and form an inversion layer. It was colder and windier than yesterday, though. Yesterday, and in fact for weeks before, the weather had been abnormally clear and dry. The consequences were visible: Pico de Orizaba, usually white-capped, was fading to granite gray. That, everyone had to admit, was not a good thing—as with all deglaciation worldwide, Pico's diminishing snow coverage was probably influenced by climate change—but they'd take it.

"Good news," David said. "We're focused."

Ten minutes to go. Everyone stood around the control desk, waiting. Shep pointed to one of David's monitors. "That cloud looks bad," he said. "You know, what's my concern is icing. If we stay in the fog and it gets cold, we could have an ice situation."

"Yes, we are pushing it," David said.

"What's the temperature?" Shep asked.

"Approaching the dew point," Gopal said.

"Not good," Shep said. "We could get some frost."

Within a few minutes, the weather had begun to shift. "You know, I don't want to be the one to say it, but it looks like things are clearing a little bit," Shep said.

On a digital clock on a large flat-screen monitor, the seconds ticked by. As 1 A.M. local time approached, Gopal counted down. "Four, three, two, one, zero. Blastoff."

David swiveled in his office chair, clapped his hands on his knees, and said, "Let's celebrate." He stood and walked around the room distributing sesame-covered cashews.

25

BLACK HOLE INITIATIVE
CAMBRIDGE, MASSACHUSETTS
APRIL 4, 2017

Shep always wondered where he'd be when the Event Horizon Telescope finally lit up. On Mauna Kea, where it all began? In Mexico, with an eye-level view of a neighboring stratovolcano, water vapor tendriling over the summit in the pre-observation gloaming? In Chile, in a spotless control room, commanding sixty-odd synchronized dishes on a sterile red plateau seven thousand feet above? Instead, he found himself in Cambridge, at the Black Hole Initiative.

Cold rain pecked the windows. Outside, the trees were covered in buds, primed for spring. Inside, it was bright and warm—hot, in fact. Ten humming computers and at least as many people were crammed into the office Shep shared with Andy Strominger, Peter Galison, and Ramesh Narayan. None of them used the room very often, because they all had bigger, nicer offices elsewhere, so for the observation, they converted it into a makeshift command center. Shep sat at a conference table covered with laptops, cellphones, and one black landline phone. In one corner of the room, a spherical black webcam pointed at the conference table.

On the north wall, a large monitor displayed satellite weather data. Beside it, on a smaller monitor, columns of clocks ticked off local time at each telescope. On a large whiteboard on the wall opposite the window, someone had drawn a chart in green, red, and blue dry-erase marker:

	WEATHER	TECHNICAL	NIGHT OUTLOOK	MULTI-DAY OUTLOOK	LOCAL WISDOM
LMT					
SPT					
SMT					
ALMA					
APEX					
SMA					
JCMT					
PICO					

On the lower right-hand corner of the whiteboard, in green marker:

4 pm: G/NG

A four o'clock each afternoon, Eastern Daylight Time, Shep would make the decision for the night: go, or no go. At that point, even if everything went to hell, there was no turning back.

It was 2 P.M. on the first day of the ten-day window. The Event Horizon Telescope Collaboration had been awarded sixty-five hours of ALMA time, which they would use in five bursts during the ten-day window. In addition to Sagittarius A* and M87, they were obliged to observe another ten quasars and black holes, the combined wishes of dozens of scientists. Vincent Fish had taken this list, considered the constraints imposed by the rotation of the Earth, the position of the sun, the location of the targets on the sky, the speed with which the different telescopes could slew between sources, and a seemingly endless list of other variables, and organized it into four "tracks," each one a sequence of about a hundred minutes-long scans. The tracks were written on a blackboard in the back of the room:

EDT

Track A #4	19:25-> 11:18
Track B #2	20:46-> 12:14
Track C #3	00:01-> 16:42
Track D #1	18:31-> 13:07

First up was track D. It didn't include the prime events—the best chances of getting an image of Sagittarius A* and M87—so it was nobody's favorite. "I'm looking at this schedule and wondering, why?" Shep said. It was too late to rethink the tracks. They had to be done. The question was whether to run tonight. In Chile, Arizona, and Spain, the skies were clear and still. The weather in Hawaii was extraordinary. The big variables were the conditions, atmospheric and technical, in Mexico and at the South Pole. Strong wind in both places threatened to turn the big primary reflectors of the Large Millimeter Telescope and the South Pole Telescope into mainsails. The forecast for the LMT said snow was on the way. And last night, the SPT was having trouble seeing easy-to-see sources. The telescope might just need to be rebooted, and if so, it would be running normally by the time the observation began. But if rebooting didn't work, the winter staff would have to work through a long troubleshooting list, and there was no telling how long that would take. Fortunately, tonight the SPT was the least essential telescope. The night's main target, M87, is in the Northern Hemisphere, so the

SPT, posted as it is at the bottom of the world, couldn't see it anyway. The real worry was that the SPT might still be offline tomorrow and beyond.

"Dan says don't wait on SPT," said Feryal Ozel. She and Dimitrios Psaltis had moved to Boston for a yearlong sabbatical at the Black Hole Initiative; Dimitrios, who now sat across the conference room table from her, had taken on the role of project scientist for the EHT.

"I'm more concerned about LMT, frankly," Shep said. For some reason, the LMT's maser was slightly out of tune.

"What's the power spectrum of maser noise?" Michael Johnson asked.

"It's one over f squared close in, then it changes," Shep said. He stood up, walked to a blackboard on the south wall, and worked through some equations. Then he and Jim Moran—wise Jim Moran, one of Shep's early mentors, here, twenty-five years after Shep showed up on Haystack's doorstep, for his mentee's hard-won experiment—looked at some plots on Shep's laptop. They decided that the problem was negligible. They'd have to fix it, but not today.

Shep received a text message from Gopal and read it to the group. "Gopal says, 'I measured the phase noise Joe. Maximum crisis level'—I think he's dictating here."

The signals from Mexico were garbled, but they had solid information from most of the other telescopes. They decided to update the whiteboard. Feryal walked up to it and uncapped a dry-erase marker, and Dimitrios read updates from the rest of the telescopes—APEX, technically ready, weather excellent; SMA, ready; LMT, weather looks good, windy but probably within the limits, and a technical go, they say.

"Any maser problems?" Feryal asked.

"Nothing that we're going to deal with right now," Shep said.

If they weren't going to worry about the LMT's maser, then the decision depended on the weather. Shep wanted to see a water-vapor map, so they pulled up a satellite feed on the big monitor. Time-lapse images of whorls of cloud sweeping over Mexico appeared on the screen. "LMT looks good to me," he said. Next, he asked to see the satellite water-vapor map for the American Southwest. He was worried about the weather at the SMT, on Mount Graham, in Arizona. The NOAA satellite imagery showed clouds smearing themselves across the state. Yet the local forecast, which he had pulled up on his laptop, said Mount Graham should be clear and blustery. "I don't know if I believe that water-vapor map," Shep said.

They went around the table debating the night's decision. They used to wonder how they'd ever get good weather simultaneously at eight telescopes on four continents, but tonight conditions were suspiciously good. It was only the first night of the window, so they could always wait and hope for things to get even better. But waiting would be risky. The weather in Mexico was likely to get worse. Besides, no one was all that excited about the track they were going to run tonight anyway. After this, Sagittarius A* was up, and the stakes got much higher. "There's something to be said for triggering the first night just to predisaster things," Shep said.

"Technically all are a go except South Pole," Feryal said. "The summary is SPT might not participate in tonight's campaign. And SMT and LMT might have a little bit of bad weather."

"This strikes me as a good set of weather," Vincent said.

After a few more minutes of repeating themselves, they reached an unspoken consensus.

"I think we should call it a go," Shep said. "So let it be written, so let it be done." He read aloud as he typed the all-go message into the Slack channel, an online chatting service, they were using as the night's primary line of communication: "Decision for 5 April: GO for VLBI. . . . This is NOT a drill."

"Don't say that," Feryal said.

Two and a half hours later, the crowd in the command center had thinned. It was dark and rainy outside, bright and quiet inside. Shep and Jason SooHoo were back at the conference table as the start of the schedule approached. "Twenty seconds until we hit," said Jason.

Shep turned to Jason. "Count down from five."

"Really?"

"Absolutely," Shep said. "I want to hear it."

"Okay," Jason said. "Five. Four. Three. Two. One. All right, things should be recording."

At 6:31 P.M. Eastern Daylight Time, the Event Horizon Telescope trained its virtual lens for the first time on an actual science target. First up was OJ287, a pair of supermassive black holes three and a half billion light-years away. The larger of the two was one of the biggest black holes ever discovered, with the mass of eighteen billion suns.

The channel they were using on Slack turned cautiously exuberant:

6:43 pm. lindyblackburn: LMT recording, scan_checks OK, still working to get point/focus up during daylight so we are not on source.

7:41 pm. sdoeleman: We see that Pico Valeta, APEX, SMT have reported things operating normally. When other sites get a chance, please confirm the schedules have started (except JCMT, SMA which are not due to start until 0100 UT)

8:10 pm. tkrichbaum: Pico is going well

8:50 pm. gbcrew: ALMA has observed all scans so far

8:50 pm. sdoeleman: XLN!

8:56 pm. jono: SMA is ready to go, but Oj287 is just below elevation limit at start of scan. Tau 0.07, phase stability excellent. . . . Luck favors the well prepared.

9:19. remo: Sorry. JCMT has been on source and observing the schedule as expected.

Hours passed. Jonathan Weintroub was running the operation in Hawaii, and things were so smooth and calm that he let the crew end the night from the satellite SMA control room at the Hale Pohaku base camp.

12:04 am. jono: SMA team is moving to HP and will monitor observation in shifts from there, with Miriam, operator. 20 minutes drive time if something breaks. So far everything has been so smooth it's downright frightening.

The sun rose over Spain, and the easternmost station called it a night.

12:38 am. hfalcke: With the exception of some (in the end minor) com-

puter issues in between, everything went smoothly. We will be wrapping up in half an hour.

1:19 am. gopal: Nice work pico!

A little before 5 A.M. Cambridge time, someone mentioned that it was Dan Marrone's birthday, and the winter crew at the South Pole, who until now had been silent, joined the conversation.

4:58 am. dpmarrone: Hi, thanks. this is not my first birthday spent observing Sgr A. All the others were on MK, however.*

5:00 am. SPT Daniel: Should have wintered at south pole for once ;)

5:01 am. gopal: Wow, SPT speaks! Hi SPT Daniel, how are things?

5:04 am. SPT Daniel: Had 5 sleepless days challenging Murphy's law as hard as we could muster. Now we are slowly converting to a well working ready/ck/ system, until Murphy hits again. Last he hit 60 minutes ago. ;) Full of hope!

At 6:59 A.M., Dan Marrone relayed news that twenty-three-knot winds at the South Pole might cause trouble there—"SPT is a big ungainly sail," he explained—but the bigger news was that the South Pole Telescope was running at all. The reboot worked.

The LMT, meanwhile, was experiencing a new problem: the telescope kept shuddering to a halt in the middle of scans. These "e-stops" were an emergency response to a dip in power flowing to the telescope. After each emergency stop they rebooted the telescope and finished the night, but they'd have to find the source of the problem before tomorrow's run.

A night of observing usually ends when night ends. Track D, however, included "daytime sources," so they kept going until almost noon . . . which was around the time Shep and the rest of the command center staff reconvened.

And by 2 P.M. Cambridge time the second day, they were back at it. The stakes were higher today, because the next track included the first real crack at Sagittarius A*. When everyone had settled in at command center, the discussion turned to the situation in Mexico. Once again, the problems there were both technological—they had to find a way to prevent these emergency stops—and meteorological: the forecast called for afternoon rain and thunderstorms. At the summit, rain and thunderstorms meant ice on the dish. And if the dish got covered in ice, they'd have to stop observing until the morning, when the sun would melt the ice away.

A voice spoke over the black speakerphone. "Hi guys, this is Lindy from LMT." He had just woken up.

"Here's the situation," Shep said. "The weather looks good everywhere. We'd kinda like to considering firing off, but everything comes down to the LMT. The first big issue is the e-stop. We can't trigger if that's gonna be bedeviling in the middle of the run. So we need some idea of what's happening other than 'someone's working on it.' Someone's gonna have to wake someone up, or call, or push, or something, but we need more information than we have."

"I'll talk to David later when we find him," Lindy said. "We were moving in slower slew mode earlier and there were some e-stops, but not as many."

"Okay, so what is the decision tree here, what are the options?" Shep said. "If even when you're slowly slewing there are some e-stops, it seems like the morning time was the problem. Maybe it's all of Mexico waking up, or the day crew using power tools. But who's trying to figure this out?"

"I think Gopal is leaving it up to Kamal in telescope engineering."

Shep picked up his iPhone, called Kamal, put the phone on speaker, and set it on the table. The emergency stops, Kamal explained, happened when the input power fell below 380 volts. It was happening at random times, but more so when the day crew showed up. He figured that for some reason, the power substation wasn't delivering enough voltage. He reduced the emergency-stop threshold to 370 volts, and he didn't exceed that level in any of the tests he ran this morning. He also asked the site manager to turn off "all other things that draw power—the heaters in the basement, other experiments."

"Kamal, this is Feryal in command center. We have been talking about slewing less often. Would that help?"

Kamal said that the difference between tracking and slewing was only about 15 volts, which is a small contribution to the voltage drop, so slewing less often wouldn't help. What would help is turning off every unnecessary device. "For example," Kamal said, "there's an electric heater in the kitchen."

"So Kamal," Shep said, "it sounds like if we freeze the people at the LMT and make them miserable, we can continue with the experiment."

"No, I'm not saying we have to freeze people," Kamal said. "I think changing the interlock threshold helps."

David Sánchez came on the line from LMT base camp. "I'm more concerned now about the weather than any interlock problem," he said. "Our weather is getting worse quickly."

"David, can you tell us more about the weather?" Shep asked. "On the radar map we're seeing that it's clear, but another site says snow showers, so it's hard to judge." The "other site" was a mountaineering website that Gopal had sent earlier that day, which Shep judged an improvement over searching Weather Underground for the forecast for Atzitzintla. "What is your best guess on ice forming, snow, et cetera?"

"I estimate a fifty percent probability of having bad weather at the beginning of the evening," David said. "That will prevent joining at the beginning, but we can probably join later unless ice forms on the surface."

"What's the probability of ice forming on the surface?" Shep asked.

"I'll say fifty," David said.

Kamal left the call, and Gopal and Aleks Popstefanija dialed in.

"Gopal and company, we've gotten the go-ahead from Kamal for the e-stops," Shep said. "Now we're dealing with weather."

"What I can tell you is we've been looking at the weather forecasts," Gopal said, "and one indicated snow showers, thunderstorms in Serdan. But we can't tell what it means for us at the site."

"We know the weather usually improves later in the day at the

LMT," Dimitrios said. "And we know the weather deteriorates over the next several days." If they passed on tonight's Sagittarius A* observation, they'd be betting that several days from now, the weather would be at least as good as it was tonight in Hawaii, Arizona, Chile, and Spain, and at the South Pole, *and* that things would have cleared up in Mexico as well. They would be coming uncomfortably close to the end of the observation window. If they lost that bet, they might miss their chance entirely.

Vincent had joined the conversation via webcam. "There's a way to make a fifty-fifty assessment," he said from a wall-mounted screen. "A very scientific instrument." He held a quarter up to the camera.

"My advice is, trigger," David said—go for it. "I don't want to cancel the night because LMT had a chance of bad weather."

"It's worse than that," Shep joked. "I'm going to say it's canceled because David said there was a fifty percent chance of bad weather. I'm going to post it on the website."

At 4:05 P.M., five minutes after the decision deadline, they were still agonizing. David Sánchez was going to give them a weather update by 6 P.M. Cambridge time. But they couldn't wait another two hours to make the call, because the people at the other telescopes needed to know what was happening—whether they should try to crash in a little more sleep before pulling another all-nighter, or whether they should let other astronomers have their telescope for the night. A big concern was wasting ALMA time. If they went for it and the weather in Mexico turned bad, had they wasted the most precious resource in radio astronomy—a night of time on ALMA? Or could they give the time back to ALMA and use it later? Based on ALMA's long history of hard-ass behavior, there was little reason to think they'd be flexible. Shep called Chile and asked anyway. He received a huge and delightful surprise: ALMA was flexible. The decision, now, was easy.

"So let it be written, so let it be done," Shep said.

At 4:40 P.M., he typed the night's instructions into Slack, reading aloud as he wrote.

> The decision for 6 april UT is A GO for VLBI the schedule is e17b06 version 10 . . . NOTE: start time of the schedule 00:46 UT April 6. This is NOT a drill.

"You said that last night," Feryal said.

• • •

They ran so hard the first three days that by the fourth, Shep faced mutiny. Each day, the forecast in Mexico was mixed, but each night, in a turnaround just short of a Full Moses, the clouds parted. By day four, Shep was like a gambler on a roll. The telescope crews, however, felt like Navy SEAL recruits on day four of Hell Week. When they realized Shep was thinking of ordering another consecutive night of observations, they simply said, "No."

When the campaign concluded on the morning of April 11, they had recorded more than sixty-five hours of data. They'd had good fortune all week. They'd suffered no catastrophic failures. When they shipped the 1,024 eight-terabyte hard drives containing the observation's harvest to Haystack and the Max Planck Institute for Radio Astronomy for correlation, the drives all arrived in good condition. The correlator operators dove into the noise in search of signal, adjusting for the drift of the atomic clocks and the wobbles of Earth and tiny uncertainties in the positions of the telescopes. They stalked abstract mathematical spaces for correlations. And one by one, they found them. Within a month, fringes were popping out of the correlator. Every thread of the web was intact. Because they didn't want to raise false hopes or encourage speculation, the collaborators were sworn to secrecy.

By the time Shep could take a proper vacation, four months after the observing run, they knew that, for the most part, everything had worked. But they had months of calibration work ahead. Shep rarely thought, *Hey, we might have one of the greatest achievements in the history of astronomy in the bag!* The daily chores—correlating, calibrating, studying plots, correcting errors, writing grants, and, of course, preparing for the next year's observing run—was too relentless and familiar to allow a sense of triumph. Plus, neither he nor anyone else involved knew what they had. What had they seen? What would they be able to show the world?

• • •

On a Monday morning in August, Shep found himself not thinking about these things for a change. He was stressed about the fog, though. He and the family had come back to Oregon for a two-week vacation. If all went well, today would be the highlight.

They were standing on the beach in Lincoln City, waiting for the first total eclipse of the sun to strike the Lower 48 since that day back in 1979, when Shep and Lane and Nels drove the Chinook out to Goldendale. The moon had already begun to overtake the sun, but through the fog and protective eclipse glasses, the sight was underwhelming—a blurry orange crescent. At the moment of totality, though, things changed. The weather gods weren't necessarily to thank for this one. The fog stayed in place, but the image burned through.

The beach got dark. People cheered and screamed. Shep, Elissa, and the kids took off their eclipse glasses and stared naked-eyed at a black circle surrounded by a ring of fire. Around the ring, starfire sprayed into space. Red fireballs bubbled up from the surface. Like everyone else on the beach that day, Shep pulled out an iPhone, pointed it at the sky, and took a picture.

EPILOGUE

A few weeks after the eclipse, Shep called his doctor to ask whether a pulse rate of thirty-five beats per minute was something to worry about. Pretty soon, emergency-room doctors were gluing electrodes to his chest. They kept him in the hospital for a week, administering tests, finding no obvious cause for the high blood pressure that had suppressed his heart rate to a slow throb. Like any good scientist, Shep was cautious about drawing causal inferences. He'd admit that he had experienced a lot of stress in the previous five years. It was true that sometimes he didn't sleep a lot. But who's to say stress caused this medical emergency? Shep would point out that he was fifty years old—no longer a kid. These things happen, he'd say. So he checked out of the hospital, quit eating salt, and went back to searching for the shadow of a black hole in the vast pool of data they had collected earlier that year.

Since late spring, when the first crates of hard drives from the April observation arrived in Boston and Bonn for correlation, the astronomers had been cautiously, gradually processing the data. They had designed a lengthy, formal, white-glove process loaded with checkpoints and safety valves to make sure they absolutely did not screw this up.

Every instrument has imperfections. The first step was finding the Event Horizon Telescope's. They ran raw data from the telescopes through the correlators until they had a good understanding of the virtual telescope's capabilities and a feel for its eccentricities. Then the Calibration and Error Analysis working group took over, analyzing their observations of well-understood quasars. Next came review meetings at which the astronomers would spend a couple days puzzling over little oddities in the data and then work up a list of things that still needed to be fixed. Then they'd send everything back to the correlator for another pass.

Through this process, they learned that during the observation much had gone right, and a fair amount had gone wrong. Sometimes a recorder stopped recording in the middle of a scan. Sometimes a little light of one polarization would "leak" into the other polarization. Most of the time, however, they found they could correct for these errors. This was all normal, all routine science. But it took time.

By the end of the year, all anyone could say for certain was that the experiment had worked—no small accomplishment—but they still had no idea what they had seen. They still hadn't correlated the Sagittarius A* and M87 data. They were still perfecting the pipeline. And, after all, they still didn't have all the data: the crew at the South Pole Telescope couldn't ship their data packs until November. When summer came to Antarctica, the crew at the South Pole Telescope took the hard drives from the spring observation out of storage, packed them into a wooden crate, loaded them onto a military aircraft, and sent them on the long journey to the correlator. On December 13, 2017, a FedEx truck pulled into the Haystack Observatory parking lot carrying that crate. The astronomers placed half of the black rectangular hard-disk modules on shelves in the correlator room along with the rest of the data packs from the 2017 run, and shipped the other half to Bonn for correlation at the Max Planck Institute for Radio Astronomy. The correlator engineers then set about adding the data they contained to that gathered by the other seven telescopes.

By early 2018, the astronomers were contending with competing priorities. They were still refining the data pipeline; the South Pole data still hadn't been fully correlated and there were weird calibration issues to work out there. They also had to prepare for the next observation, scheduled for April 2018, in which they would double the bandwidth at all sites and add new telescopes, including a new station in Greenland. They argued. What was more important: fielding a bigger, more powerful Earth-size telescope in 2018, or finishing the job they'd started the year before? There simply weren't enough people to do everything that needed to be done. They weren't the only ones who wanted to know what was in the 2017 data. The institutes and funding agencies that had footed the bill for the experiment were leaning on Shep, pushing for results, already making preliminary plans for how to announce them. But it was too early. In February, the astronomers made the agonizing decision to take a break from working on the 2017 data long enough to carry out the 2018 observing run.

And so, once again, like migrating birds they made their annual trips to their telescopes. This time out they discovered another contingency—another unpredictable variable besides technical issues and weather—that they'd have to keep in mind in the future: bandits. On Monday, April 23, the second-shift night crew at the Large Millimeter Telescope was driving from base camp to the summit when two unmarked trucks, each loaded with five armed men, stopped them for questioning. Nothing like this had ever happened before at the LMT, but the region was now home to conflict between police and gangs who were stealing gas from the cross-country pipeline that ran through the area. The armed men let the astronomers pass, and no one was harmed, but the incident was considered a major security threat. David Hughes sent everyone back to Ciudad Serdan or Puebla. Shep pulled the Large Millimeter Telescope out of the rest of the observing run, leaving a hole in the Event Horizon Telescope, and nearly guaranteeing that until they get another chance in 2019, that first big observation in 2017 remains their best chance at seeing the shadow of a black hole.

On June 5, 2018, the astronomers formally released the final calibrated Sagittarius A* and M87 data to the four working groups tasked with making images. To avoid poisoning each other's minds—so that no one could accidentally nudge another group into seeing a black-hole shadow that wasn't really in the data—the groups worked in isolation and secrecy, making images using different algorithms and techniques, trying hard to falsify anything that looked too sharp, too clean, too likely to be the product of wishful thinking.

The world will know what the astronomers have seen after the imaging teams converge on a single, unmistakably real picture; after the astronomers have thought hard about what it means; and after their results have survived the scrutiny of a peer-reviewed scientific journal. It's possible they'll encounter what Shep calls the nose-of-God scenario, in which an unmistakable image of the shadow of Sagittarius A* or M87 easily and quickly comes into focus. Other experiments have had such good luck. In 2015, when the scientists of the LIGO gravitational-wave observatory started analyzing their inaugural data set, they were stunned to immediately find the signature spacetime ripple of the merger of two distant stellar-mass black holes. At the other end of the spectrum of possibilities is failure: they'll see nothing.

Pure failure seems highly unlikely. The telescope worked; it saw something. The question is, what? Is it the artful crescent shape the models predict? Is it a mess? Is it something in between? We now know that with millions of dollars and the combined force of will of a couple hundred people, human beings can construct a network of machines capable of seeing what the best theories of nature predict is the universe's nearest major exit door. But are those theories right?

A rush of scientific papers released in anticipation of the Event Horizon Telescope's results showed that deciding what the shadow image tells us could be onerous and contentious. In April 2018, *Nature Astronomy* published a paper by a group of scientists including Heino Falcke, Michael Kramer, and Luciano Rezzolla, the BlackHoleCam trio. After comparing simulated images of a Kerr black hole with an exotic black hole created using an alternative theory of gravity, they concluded that "it could be extremely difficult to distinguish between black holes from different theories of gravity, thus highlighting that great caution is needed when interpreting black hole images as tests of general relativity." A couple months later, four French scientists writing in the journal *Classical and Quantum Gravity* argued that a hypothetical "non-singular" black hole without an event horizon and even a wormhole would look, to the EHT, a whole lot like the shadow of a Kerr black hole—the latter, of course, being what all mainstream thinking predicts Sagittarius A* really is. Could it be that a clear picture of the long-anticipated, Einstein-predicted shadow wouldn't rule out the possibility that Sagittarius A* is, in fact, something as weird as a wormhole?

So even a pristine, searing image of the shadow of the black hole at the center of the Milky Way won't end the story. The image and all the accompanying data will be picked apart, fussed over, and attacked from various angles. It will be interpreted in light of other recent experimental and theoretical advances. But even if no one immediately agrees on what the picture tells us, its arrival, if and when that happens, could signal the beginning of a new era—with luck, one in which people gain new traction in the long and baffling quest to understand what happens in those dark places where space-time ends.

Acknowledgments

I'm grateful to the scientists of the Event Horizon Telescope for their childlike openness to my lurking, note-taking presence over the past six years, and also for never hiring a public relations person. Thanks to Shep Doeleman, Elissa Weitzman, Lane Koniak, Nels Doeleman, Jonathan Weintroub, Robbie Singal, Rurik Primiani, Laura Vertatschitsch, Michael Johnson, Alan Rogers, Jim Moran, Colin Lonsdale, Mike Hecht, Vincent Fish, Jason SooHoo, Mike Titus, Rusen Lu, Lucy Zuryis, Dimitrios Psaltis, Feryal Ozel, Geoff Bower, Dan Marrone, Junhan Kim, Avery Broderick, Heino Falcke, Remo Tilanus, Thomas Krichbaum, Alan Roy, Fulvio Melia, David Hughes, Gopal Narayanan, David Sánchez, Jonathan León-Tavares, Gisela Ortiz, Pierre Cox, Geoff Crew, Lynn Matthews, Katie Bouman, Lindy Blackburn, Andrew Chael, Sera Markoff, Avi Loeb, Andrea Ghez, Daryl Haggard, Fred Baganoff, Ken Kellermann, Ron Ekers, Miller Goss, Bruce Balick, Steve Giddings, Andrew Strominger, Alex Lupsasca, Janna Levin, Priyamvada Natarajan, and others I'm sure I'm forgetting to name. Donald Lynden-Bell and Joseph Polchinski departed this world before it was time to compile this list. I'm fortunate to have spoken to them while they were still here.

Thanks to Larry Weissman and Sascha Alper for their expert representation, and to Hilary Redmon and Dan Halpern at Ecco for taking a chance on this project in 2013, when the Event Horizon Telescope was far from a sure thing. Thanks to Denise Oswald for taking this project on enthusiastically, and to Emma Janaskie for her unfailing helpfulness.

I'm grateful to my colleagues at two magazines: from the *Popular Science* days, when this project was just getting started, Mark Jannot, Luke Mitchell, Jake Ward, and Cliff Ransom; at *Scientific American*, Mariette DiChristina, Fred Guterl, Curtis Brainard, Christi Keller, Michael Mrak, Michael Lemonick, Dean Visser, Lee Billings, Clara Moskowitz, Kate Wong, Jen Schwartz, Michael Moyer, Robin Lloyd, and George Musser. I'm indebted to Dan Baum for a well-timed lecture on the virtues of narrative nonfiction written in the close third person, organized chronologically.

A generous grant from the Alfred P. Sloan Foundation pushed this book over the finish line. Thanks to Doron Weber and Eliza French at the Sloan Foundation for their vision and support. Sloan funds enabled me to enlist a talented trio of journalists to help with the final detail work. Matt Mahoney fact-checked large portions of the book, but all remaining mistakes are mine. Andrea Marks helped with research, compiling front and end matter, and offered sharp comments on the manuscript at several stages. Katie Peek, astrophysicist turned journalist and artist, made gorgeous maps and diagrams for the book. I owe Katie and Josh Peek extra thanks for tipping me off in January 2012 or so to the existence of the Event Horizon Telescope. Christian Debenedetti, Abe Streep, John Gilbreth, Josh Dean, Gabe Sherman, Jen Stahl, Adrianne Cohen, Catherine Price, Andrew Blum, and Liesl Schillinger all provided advice and mental support over the years. Jamie and Michelle Hough, Ellen Garrison, and my mom, Ann Banks, helped with childcare and who knows what else when travel schedules got complicated and deadlines got tight.

I don't think it's possible to adequately thank my wife, Leigh, and our daughter, Sylvia, for their patience and support. I started this project when Sylvia was four months old. Now she's four feet tall. Thank you, Sylvia, for putting up with your distracted, exhausted dad. This book is for you.

Notes

Although I removed myself from the narrative, I witnessed many of the events depicted in this book, particularly those that happened in 2012 or later. Scenes I did not witness I have re-created using standard journalistic methods—by interviewing the people involved and consulting published accounts, transcripts, weather records, photos, videos, and maps.

1

3 **a cold February morning**: The description of the eclipse of 1979 is based on interviews with Shep Doeleman and his family, contemporaneous news accounts, weather records, and photos and videos from the event.

2

5 **a bit of trouble**: The account of Eddington's background and the events leading up to the 1919 eclipse expedition relies heavily on Matthew Stanley, "'An Expedition to Heal the Wounds of War': The 1919 Eclipse and Eddington as Quaker Adventurer," *Isis* 94, no. 1 (2003).

6 **timeless hallmarks**: Albert Einstein, *The Collected Papers of Albert Einstein*, trans. Ann Hentschel, vol. 9, *The Berlin Years: Correspondence, January 1919–April 1920* (Princeton, NJ: Princeton University Press, 2004).

6 **a dog named Nipper**: Letters from A. S. Eddington to Sarah Ann Eddington and Winifred Eddington: Madeira and the Eclipse at Principe, Papers of Sir Arthur Eddington, Janus, Trinity/EDDN/A4, 317, https://janus.lib.cam.ac.uk/db/node.xsp?id=EAD/GBR/0016/EDDN/A4.

6 **official report on the expedition**: F. W. Dyson, A. S. Eddington, and C. Davidson, "A Determination of the Deflection of Light by the Sun's Gravitational Field, from Observations Made at the Total Eclipse of May 29, 1919," *Philosophical Transactions of the Royal Society of London A* 220 (January 1, 1920).

6 **eclipse expedition**: See Stanley, 2003.

7 **in *The ABC of Relativity***: Bertrand Russell, *The ABC of Relativity* (London: K. Paul, Trench, Trubner, 1931), p. 24.

8 **"On the Electrodynamics of Moving Bodies"**: Albert Einstein, "Zur Elektrodynamik bewegter Körper," Annalen der Physik 322 (10): 891–921, and *The Principle of Relativity: Original Papers by A. Einstein and H. Minkowski*, University of Calcutta, 1920, pp. 30–63, http://ebook.lib.hku.hk/CADAL/B31400541/.

9 **are, in fact, flexible**: You could fill an aircraft hangar with books and articles on special and general relativity. I relied primarily on Russell, Eddington, and the undergraduate-level textbook *Exploring Black Holes* by Edwin F. Taylor and John Archibald Wheeler.

9 **in a famous lecture in 1908**: Hermann Minkowski, "Raum Und Zeit," *Physicalische Zeitschrift* 10 (1909).

10 **in the external world**: Arthur Stanley Eddington, *Space, Time and Gravitation: An Outline of the General Relativity Theory* (Cambridge: Cambridge University Press, 1920).

10 **his "happiest thought"**: Walter Isaacson's biography of Einstein is as good a source as any for this widely recorded anecdote. Walter Isaacson, *Einstein: His Life and His Universe* (New York: Simon & Schuster, 2008), p. 145.

10 **"together in the unity and wholeness"**: Quoted as it appears in *Exploring Black Holes*.

11 **Near the end of his magnum opus**: *Philosophiae Naturalis Principia Mathematica.* Isaac Newton, 1642–1727, *Principia.* English: *Sir Isaac Newton's Mathematical Principles of Natural Philosophy and His System of the World,* translated into English by Andrew Motte in 1729. The translations revised, and supplied with a historical and explanatory appendix, by Florian Cajori (Berkeley, California: University of California Press, 1934), pp. 371–72.

11 **"Spacetime tells matter"**: John Archibald Wheeler and Kenneth William Ford, *Geons, Black Holes, and Quantum Foam: A Life in Physics,* 1st ed. (New York: Norton, 1998), p. 235.

12 **a small discrepancy**: This is probably a good place to address the accusation that Eddington threw his experiment in Einstein's favor. Short version: he probably didn't. See Stanley, 2003.

12 **Hermann Weyl wrote**: Hermann Weyl, *Space, Time, Matter.* Translated from the German by Henry L. Brose. (London: Metheun & Co. Ltd., 1922), p. iii.

3

19 **from MIT's Rad Lab**: Alan R. Whitney, Colin J. Lonsdale, and Vincent L. Fish, "Insights into the Universe: Astronomy with Haystack's Radio Telescope," *Lincoln Laboratory Journal* 21, no. 1 (2014).

19 **do astronomy:** Ibid.

20 **as Einstein's theory predicted**: Ibid.

4

23 **"your land of ideas"**: Jeffrey Crelinsten, *Einstein's Jury: The Race to Test Relativity* (Princeton, NJ: Princeton University Press, 2006).

24 **when Subrahmanyan Chandrasekhar**: See Arthur I. Miller, *Empire of the Stars: Obsession, Friendship, and Betrayal in the Quest for Black Holes* (Boston: Houghton Mifflin, 2005).

25 **"will continue indefinitely"**: J. R. Oppenheimer and H. Snyder, "On Continued Gravitational Contraction," *Physical Review* 56 (1939).

25 **"since the time of Galileo"**: Stephen Hawking and W. Israel, *Three Hundred Years of Gravitation* (Cambridge and New York: Cambridge University Press, 1987).

25 **cover of *Time***: *Time*, March 11, 1966.

27 **wearing a sling**: See Israel, "Dark Stars: The Evolution of an Idea," in Hawking and Israel, *Three Hundred Years of Gravitation*, p. 245.

27 **"dimmer millisecond by millisecond"**: Kip S. Thorne, *Black Holes and Time Warps: Einstein's Outrageous Legacy* (New York: W.W. Norton, 1994), p. 256.

27 **"the external constraints"**: Israel, "Dark Stars," p. 259.

27 **Wheeler gave them**: Wheeler gave credit to an audience member for suggesting the name "black hole," but he embraced the term and is widely credited for the coinage.

28 **waterfalls**: Andrew Hamilton, "A Black Hole Is a Waterfall of Space," http://jila.colorado.edu/~ajsh/insidebh/waterfall.html, retrieved February 21, 2018.

28 **"a perfect unidirectional membrane"**: David Finkelstein, "Past-Future Asymmetry of the Gravitational Field of a Point Particle," *Physical Review* 110, no. 4 (1958).

29 **"In my entire scientific life"**: Fulvio Melia, *Cracking the Einstein Code: Relativity and the Birth of Black Hole Physics* (Chicago: University of Chicago Press, 2009), p. 76.

29 **Roger Penrose used**: Roger Penrose, "Gravitational Collapse: The Role of General Relativity," *Nuovo Cimento Rivista Serie* 1 (1969).

30 **what makes quasars shine**: The Penrose process has since been supplanted by more complex models, but the basic picture still holds.

31 ***was* its entropy**: Technically, as Bekenstein wrote, "the black-hole entropy is equal to the ratio of the black-hole area to the square of the Planck length times a dimensionless constant of order unity." See Jacob D. Bekenstein, "Black Holes and Entropy," *Physical Review D* 7, no. 8 (1973): 2333–46.

32 **"Black Hole Explosions?"**: S. W. Hawking, "Black Hole Explosions?," *Nature* 248 (1974).

32 **ten-billionths of a degree**: Leonard Susskind, *The Black Hole War: My Battle with Stephen Hawking to Make the World Safe for Quantum Mechanics* (New York: Little, Brown and Co, 2008), chapter 9, iBooks.

33 **John Preskill later wrote**: J. Preskill and D. V. Nanopoulos, "Do Black Holes Destroy Information?," in *Black Holes, Membranes, Wormholes and Superstrings, Proceedings of the International Symposium,* Houston, TX, January 16–18, 1992, edited by Sunny Kalara and D. V. Nanopoulos (Singapore: *World Scientific*, 1993), p. 22.

33 **They launched Geiger counters**: See Thorne, *Black Holes and Time Warps*, pp. 309–19.

33 **a song cycle**: Lyrics by Neil Peart, published on the Rush home page, https://www.rush.com/songs/cygnus-x-1-book-one-the-voyage/, retrieved February 21, 2018.

34 **Donald Lynden-Bell**: This account draws on interviews with Lynden-Bell and Ron Ekers. See also Donald Lynden-Bell, "Searching for Insight," *Annual Review of Astronomy and Astrophysics* 48 (2010); D. Lynden-Bell, "Galactic Nuclei as Collapsed Old Quasars," *Nature* 223 (1969); and D. Lynden-Bell and M. J. Rees, "On Quasars, Dust and the Galactic Centre," *Monthly Notices of the Royal Astronomical Society* 152 (1971).

35 **they found odd specks of radiation**: R. D. Ekers and D. Lynden-Bell, "High Resolution Observations of the Galactic Center at 5 Ghz," *Astrophysical Letters* 9 (1971).

35 **announcing their discovery**: B. Balick and R. L. Brown, "Intense Sub-Arcsecond Structure in the Galactic Center," *Astrophysical Journal* 194 (1974). For a more detailed history, see W. M. Goss, Robert L. Brown, and K. Y. Lo, "The Discovery of Sgr A*," *Astronomische Nachrichten Supplement* 324 (2003).

5

38 **a group led by Charles Townes**: R. Genzel and C. H. Townes, "Physical Conditions, Dynamics, and Mass Distribution in the Center of the Galaxy," *Annual Review of Astronomy and Astrophysics* 25 (1987).

38 ***couldn't* be a black hole**: See R. H. Sanders, "The Case Against a Massive Black Hole at the Galactic Centre," *Nature* 359 (1992).

40 **Karl Guthe Jansky**: See Thorne, *Black Holes and Time Warps,* pp. 322–45, for a longer retelling of the story of Jansky and Reber.

40 **"a very steady continuous interference"**: W. A. Imbriale, "Introduction to 'Electrical Disturbances Apparently of Extraterrestrial Origin,'" *Proceedings of the IEEE* 86, no. 7 (1998).

41 **Jansky's discovery**: Karl G. Jansky, "Electrical Disturbances Apparently of Extraterrestrial Origin," *Proceedings of the IRE* 21, no. 10 (October 1993):

1387–98, reprinted in 1998 by IEEE and available at https://www.ieee.org/documents/jansky.pdf.

41 **Grote Reber**: Thorne, *Black Holes and Time Warps,* pp. 324–27.

42 **chicken wire**: K. I. Kellermann and J. M. Moran, "The Development of High-Resolution Imaging in Radio Astronomy," *Annual Review of Astronomy and Astrophysics* 39 (2001).

42 **the first radio interferometers**: For this brief history of the development of VLBI, I relied primarily on articles by and interviews with Jim Moran and Ken Kellermann. See Kellermann and Moran, "The Development of High-Resolution Imaging in Radio Astronomy," and Kellerman's article "Intercontinental Radio Astronomy," *Scientific American,* February 1972.

42 **led by Martin Ryle**: Malcolm Longair, "A Brief History of Radio Astronomy in Cambridge," University of Cambridge, https://www.astro.phy.cam.ac.uk/about/history, accessed February 21, 2018.

43 **All succeeded by 1967**: Kellermann and Moran, "The Development of High-Resolution Imaging in Radio Astronomy."

43 **ten thousand kilometers**: Ibid.

45 **Lovecraft's short story**: H. P. Lovecraft, *The Complete Fiction of H. P. Lovecraft* (London: Chartwell Books, 2016).

6

50 **an absent sun**: Reinhard Genzel and Andreas Eckart, "The Galactic Center Black Hole," and "Mid-Infrared Imaging of the Central Parsec with Keck, Angela Cotera et al.," Central Parsecs of the Galaxy, ASP Conference Series, vol. 186, edited by Heino Falcke et al.

51 **Krichbaum told the room**: A transcript of this discussion can be found in Anton Zensus and Heino Falcke, "Can VLBI Constrain the Size and Structure of SGR A*?" Zensus, J. A. and Falcke, H "The Central Parsecs of the Galaxy," ASP Conference Series, vol. 186, edited by Heino Falcke, Angela Cotera, Wolfgang J. Duschi, Fulvio Melia, and Marcia J. Rieke et al., 1999, p. 118.

51 **"observing this effect"**: James Bardeen, "Timelike and Null Geodesics in the Kerr Metric," in *Black Holes,* edited by C. DeWitt and B. S. DeWitt (New York: Gordon and Breach, 1973), p. 215.

52 **did some similar calculations**: J. P. Luminet, "Image of a Spherical Black Hole with Thin Accretion Disk," *Astronomy and Astrophysics* 75, no. 1–2 (May 1979): 228–35.

52 **a nineteenth-century French poem**: In *Les Chimères,* Paris, 1854. Translation by Jean-Pierre Luminet.

53 **The size of the shadow**: To avoid a lengthy, distracting, and ultimately unnecessary digression about the changing estimates for the mass of Sagittarius A* and the impact that had on the expected size of its

shadow, in this paragraph I've split the difference between early and later estimates of the shadow size. For the record, though, when Heino Falcke et al. wrote the 2000 shadow paper, the best estimate for the mass of Sagittarius A* was 2.6 million solar masses, which would yield a shadow of about thirty microarcseconds, which would be tough to see clearly at a wavelength of one millimeter. Within a few years, the mass estimate for Sagittarius A* was upgraded to four million solar masses, and expectations for the shadow grew accordingly.

54 **published their findings**: Heino Falcke, Fulvio Melia, and Eric Agol, "Viewing the Shadow of the Black Hole at the Galactic Center," *Astrophysical Journal Letters* 528, L13 (2000).

54 **"in the next few years"**: "First Image of a Black Hole's 'Shadow' May Be Possible Soon," Max Planck Institut für Radioastronomie press release, January 17, 1999.

7

55 **an enthusiastic article**: Erik Stokstad, "Into the Lair of the Beast." *Science* 287, no. 5450 (2000): 65–67.

55 **CNN.com**: Richard Stenger, "New Telescope as Big as Earth Itself," CNN.com, October 2, 2002, http://www.cnn.com/2002/TECH/space/10/02/radio.telescope/index.html, retrieved February 21, 2018.

55 **seven millimeters**: Geoffrey C. Bower et al., "Detection of the Intrinsic Size of Sagittarius A* Through Closure Amplitude Imaging," *Science* 304 (2004).

56 **to melt a snowflake**: The meagerness of the energies involved is built into the Jansky, the radio astronomer's unit of measure for flux density, which, roughly speaking, is the amount of power that reaches a telescope from a cosmic source. The amount of power being measured per unit of telescope in one Jansky is 0.00000000000000000000000001 watts.

59 **The results**: Zhi-Qiang Shen, K. Y. Lo, M. C. Liang, Paul T. P. Ho, and J. H. Zhao. "A Size of 1au for the Radio Source Sgr A* at the Centre of the Milky Way." *Nature* 438 (November 01, 2005): 62–64.

59 **The *New York Times***: "Astronomers Say They Are on the Verge of Seeing a Black Hole," Dennis Overbye, *The New York Times*, November 2, 2005, https://www.nytimes.com/2005/11/02/science/astronomers-say-they-are-on-the-verge-of-seeing-a-black-hole.html.

61 **going down the drain**: Ramesh Narayan et al., "Advection-Dominated Accretion Model of Sagittarius A*: Evidence for a Black Hole at the Galactic Center," *Astrophysical Journal* 492, no. 2 (1998).

61 **"hotspots"**: Avery E. Broderick and Abraham Loeb, "Imaging Bright-Spots in the Accretion Flow Near the Black Hole Horizon of Sgr A*," *Monthly Notices of the Royal Astronomical Society* 363 (2005).

65 **"whether human or divine"**: Udías Augustín, *Searching the Heavens*

and the Earth: The History of Jesuit Observatories (Dordrecht: Kluwer Academic, 2003).

8

72 **The results**: Sheperd S. Doeleman et al., "Event-Horizon-Scale Structure in the Supermassive Black Hole Candidate at the Galactic Centre," *Nature* 455 (2008).

9

80 **submission of a paper**: Sheperd Doeleman et al., "Imaging an Event Horizon: submm-VLBI of a Super Massive Black Hole," in *astro2010: The Astronomy and Astrophysics Decadal Survey* (2009).

10

90 **Mercury, Venus, Earth, and Mars**: "they'll probably stay in roughly the same orbits"—things could go another way. Astronomers have estimated that there is a 1–2 percent chance that sometime between forty million and five billion years from now, Mercury's orbit could grow so elongated that it would cross orbits with Venus. That crossing would send the inner solar system into turmoil, possibly sending Earth smashing into Mars, in which case, one of the astronomers told *New Scientist*, "all life gets extinguished immediately, and Earth glows at the temperature of a red giant star for about 1,000 years."

91 **Soviet theorists**: N. I. Shakura and R. A. Sunyaev, "Black Holes in Binary Systems. Observational Appearance," *Astronomy and Astrophysics* 24 (1973).

91 **Steven Balbus and John Hawley**: Steven A. Balbus and John F. Hawley, "A Powerful Local Shear Instability in Weakly Magnetized Disks. I-Linear Analysis. Ii-Nonlinear Evolution," *Astrophysical Journal* 376 (1991).

11

95 **Spinoza Prize**: "NWO-Spinoza Prize for Heino Falcke, Patti Valkenburg and Erik Verlinde," June 6, 2011, https://www.nwo.nl/en/news-and-events/news/2011/NWO-Spinoza+Prize+for+Heino+Falcke,+Patti+Valkenburg+and+Erik+Verlinde.html, retrieved February 21, 2018.

98 **press release**: "ERC Synergy Grant to Image Event Horizon of Black Hole," Radboud University press release, December 17, 2013, http://www.ru.nl/english/@928308/pagina/, retrieved February 20, 2018.

13

105 **"inaugurated"**: "Giant Mexican Telescope Launched," BBC.com, November 23, 2006, http://news.bbc.co.uk/2/hi/science/nature/6175446.stm, retrieved February 21, 2018.

106 **"visit of supervision"**: Juan Cervantes, "Calderón y Moreno Valle supervisan operaciones del telescopio milimétrico," Sobre-T.com, September 21, 2012, https://www.sobre-t.com/calderon-y-moreno-valle-supervisan-operaciones-del-telescopio-milimetrico/, retrieved February 21, 2018.

107 **every million years**: A hydrogen maser takes a bunch of hydrogen atoms and nudges them into the same "hyperfine" state, so they coherently emit the same rarified light—electromagnetic radiation with a wavelength of 21 centimeters and a frequency of 1,420 megahertz, or 1.42 billion oscillations per second. Circuitry amplifies this pure, focused hyperfine emission and uses it to control the output of a quartz oscillator. When the hydrogen in the cavity and the quartz oscillator are humming along in harmony, the maser is "phase locked." If, over time, they drift apart, the maser will apply a corrective voltage to the quartz oscillator.

15

121 **Black-hole complementarity**: David A. Lowe et al., "Black Hole Complementarity Versus Locality," *Physical Review D* 52, no. 12 (1995).

122 **bet John Preskill**: George Johnson, "What a Physicist Finds Obscene," *New York Times*, February 16, 1997.

122 **conceded that bet**: Jenny Hogan, "Hawking Concedes Black Hole Bet," *New Scientist*, July 21, 2004.

123 **complementarity didn't work**: Ahmed Almheiri et al., "Black Holes: Complementarity or Firewalls?," *Journal of High Energy Physics* 2 (2013).

123 **Polchinski wrote**: Joseph Polchinski, "Rings of Fire," *Scientific American*, April 2015.

124 **"I've never been so surprised"**: Dennis Overbye, "A Black Hole Mystery Wrapped in a Firewall Paradox," *New York Times*, August 12, 2013.

124 **Stephen Hawking joined**: S. W. Hawking, "Information Preservation and Weather Forecasting for Black Holes," submitted to arXiv.org January 22, 2014, https://arxiv.org/abs/1401.5761.

124 **humor piece**: Andy Borowitz, "Stephen Hawking's Blunder on Black Holes Shows Danger of Listening to Scientists, Says Bachmann," *New Yorker*, January 27, 2014, https://www.newyorker.com/humor/borowitz-report/stephen-hawkings-blunder-on-black-holes-shows-danger-of-listening-to-scientists-says-bachmann, retrieved February 21, 2018.

125 **everything exists in some place**: George Musser's book *Spooky Action at a Distance* is helpful for understanding the slippery concept of locality.

125 **a remnant**: Steven B. Giddings, "Black Holes and Massive Remnants," *Physical Review D* 46 (1992).

127 **would glisten**: Stephen B. Giddings, "Possible Observational Windows for Quantum Effects from Black Holes," *Physical Review D* 90 (2014).

16

131 **where Pluto was**: "ALMA Pinpoints Pluto to Help Guide NASA's New Horizons Spacecraft," National Radio Astronomy Observatory press release, August 5, 2014, https://public.nrao.edu/news/alma-pluto/, retrieved February 21, 2018.

17

138 **one cosmologist told the *New York Times***: Dennis Overbye, "Space Ripples Reveal Big Bang's Smoking Gun," *New York Times*, March 17, 2014.

138 **In a YouTube video**: "Stanford Professor Andrei Linde Celebrates Physics Breakthrough," Stanford University, March 17, 2014, https://www.youtube.com/watch?v=ZlfIVEy_YOA, retrieved February 21, 2018.

138 **"Blockbuster Big Bang Result May Fizzle, Rumor Suggests"**: Adrian Cho, *Science*, May 12, 2014, http://www.sciencemag.org/news/2014/05/blockbuster-big-bang-result-may-fizzle-rumor-suggests, retrieved February 21, 2018.

139 **"Stardust got in their eyes"**: Dennis Overbye, "Criticism of Study Detecting Ripples from Big Bang Continues to Expand," *New York Times*, September 22, 2014.

139 **motion microscope**: Frédo Durand, William T. Freeman, and Michael Rubinstein, "Video Microscope Reveals Movement in 'Stock-Still' Objects," *Scientific American*, January 2015.

139 **"visual microphone"**: Abe Davis et al., "The Visual Microphone: Passive Recovery of Sound from Video," SIGGRAPH 2014, http://people.csail.mit.edu/mrub/VisualMic/, retrieved February 21, 2018.

23

197 **Calabi-Yau spaces**: See Shing-Tung Yau and Steven J. Nadis, *The Shape of Inner Space: String Theory and the Geometry of the Universe's Hidden Dimensions* (New York: Basic Books, 2010).

197 **"soft hair"**: Stephen W. Hawking, Malcolm J. Perry, and Andrew Strominger, "Soft Hair on Black Holes," submitted to arXiv.org January 5, 2016, https://arxiv.org/abs/1601.00921, retrieved February 21, 2018.

199 **Their predictions**: Samuel E. Gralla, Alexandru Lupsasca, and Andrew Strominger, "Near-horizon Kerr Magnetosphere," Arxiv.com, submitted February 4, 2016, last revised May 24, 2016, https://arXiv.org/abs/1602.01833, retrieved February 21, 2018.

Glossary

Baseline The distance between a pair of antennas in a Very Long Baseline Interferometry array.

Black hole A region of space from which nothing that enters can escape.

Black-hole complementarity A theory applying particle-wave duality and the holographic principle to black holes, concluding that someone falling into a black hole would appear smeared across the event horizon to an observer, and further, that information can be rescued from black holes.

Black-hole information paradox A long-standing problem in theoretical physics that results from the prediction that black holes will eventually evaporate, destroying all information about their contents. The rules of quantum mechanics strictly forbid the destruction of information, hence the trouble.

Curved spacetime In Einstein's general theory of relativity, matter (and, equivalently, energy) distort or "curve" spacetime. The path of objects through curved spacetime creates what we experience as gravity.

Event horizon The boundary of a black hole, beyond which neither matter nor light can escape.

Firewall argument The argument, advanced in 2012 by a group of theoretical physicists led by the late Joseph Polchinski, that the event horizon of a black hole is not, in fact, empty space—it is the site of a dramatic break in spacetime that incinerates everything that hits it.

Frame-dragging The process by which a spinning black hole would pull spacetime itself around with it.

Fringes In Very Long Baseline Interferometry observations, a common detection between two antennas.

Gravitational waves Ripples in spacetime caused by the mergers of black holes and other violent events.

Holographic principle The proposal that all information about the contents of a black hole is stored on a surface just outside the event horizon.

Interferometry In radio astronomy, a method of observing with two or more geographically distant telescopes and combining the collected data to form a single master output.

Length contraction In relativity, the phenomenon in which, to an outside observer, an object traveling at close to the speed of light contracts in the direction of travel.

Metric A mathematical formula for measuring the separation between events in a given spacetime geometry.

Moore's law The prediction, made by Gordon Moore in 1965, that the number of components on an integrated circuit would double every year for the next two years, as they had for the previous decade. He later revised his prediction to say that the density of integrated circuits would double every two years.

No-hair theorem The idea, widely accepted but never proven, that black holes have no imperfections—no "hair"—and can be completely characterized by their mass, angular momentum, and electric charge.

Quantum entanglement The phenomenon in which the quantum states of two particles become inextricably linked, even when those particles are separated by great distances.

Quantum mechanics The physical theory that describes nature at the subatomic level.

Quasar Short for quasi-stellar radio sources. Originally called radio stars.

Radio astronomy The branch of astronomy whose practitioners study celestial objects by collecting light in the long-wavelength "radio" portion of the electromagnetic spectrum.

Radio galaxies Discovered by radio astronomers after World War II, fountains of radio energy emanating from patches of sky that had appeared empty to optical astronomers.

Relativity, Einstein's general theory of Albert Einstein's theory of gravity, which treats the universe as a four-dimensional spacetime continuum and gravity as the geometry, or curvature, of spacetime.

Relativity, principle of The requirement that the laws of physics take the same form in all inertial reference frames.

Sagittarius A* The supermassive black hole at the center of the Milky Way Galaxy.

Singularity An undefined point, the equivalent of dividing by zero. At the singularities at the center of black holes, the known laws of physics break down.

Spacetime A mathematical continuum consisting of three spatial coordinates and one temporal coordinate.

Submillimeter wavelengths The highest frequencies of radio light. The slice of spectrum between the microwave and the infrared.

Tau A ratio, used by radio astronomers, that measures the opacity of atmosphere to starlight.

Time dilation In relativity, the slowing of time at velocities approaching the speed of light and in strong gravitational fields.

Very Long Baseline Interferometry (VLBI) A method in which astronomers observe simultaneously with two or more geographically distant radio telescopes and then combine the collected data to form a single master output.

Selected Bibliography

BOOKS

Barrow, John D. *Cosmic Imagery: Key Images in the History of Science.* 1st Amer. ed. New York: Norton, 2008.

Begelman, Mitchell C., and Martin J. Rees. *Gravity's Fatal Attraction: Black Holes in the Universe.* Scientific American Library Series. New York: Scientific American Library, 1996. Distributed by W. H. Freeman.

Bohm, David. *Wholeness and the Implicate Order.* London and Boston: Routledge & Kegan Paul, 1981.

Davis, Joel. *Journey to the Center of Our Galaxy: A Voyage in Space and Time.* Chicago: Contemporary Books, 1991.

Eddington, Arthur Stanley. *Science and the Unseen World.* Swarthmore Lecture. New York: Macmillan, 1929.

——. *Space, Time and Gravitation: An Outline of the General Relativity Theory.* Cambridge, UK: University Press, 1920.

Einstein, Albert. *The Collected Papers of Albert Einstein.* Vol. 9, *The Berlin Years: Correspondence, January 1919–April 1920.* Translated by Ann Hentschel. Princeton, NJ: Princeton University Press, 2004.

Ferreira, Pedro G. *The Perfect Theory: A Century of Geniuses and the Battle over General Relativity.* Boston: Houghton Mifflin Harcourt, 2014.

Ferris, Timothy. *Coming of Age in the Milky Way.* 1st ed. New York: Morrow, 1988.

——. *The Whole Shebang: A State-of-the-Universe(s) Report.* New York: Simon & Schuster, 1997.

Feynman, Richard P. *Six Not-So-Easy Pieces: Einstein's Relativity, Symmetry, and Space-Time.* Reading, MA: Addison-Wesley, 1997.

Galison, Peter, Gerald James Holton, and S. S. Schweber. *Einstein for the 21st Century: His Legacy in Science, Art, and Modern Culture.* Princeton, NJ: Princeton University Press, 2008.

Guth, Alan H. *The Inflationary Universe: The Quest for a New Theory of Cosmic Origins.* Reading, MA: Addison-Wesley, 1997.

Hawking, Stephen, and W. Israel. *Three Hundred Years of Gravitation.* Cambridge and New York: Cambridge University Press, 1987.

Isaacson, Walter. *Einstein: His Life and His Universe.* New York: Simon & Schuster, 2008.

Koestler, Arthur. *The Sleepwalkers: A History of Man's Changing Vision of the Universe.* London and New York: Arkana, 1959.

Léna, Pierre, and Laurent Mugnier. *Observational Astrophysics.* Astronomy and Astrophysics Library. 3rd ed. Heidelberg and New York: Springer, 2012.

Luminet, Jean-Pierre. *Black Holes.* Cambridge and New York: Cambridge University Press, 1992.

Maudlin, Tim. *Philosophy of Physics: Space and Time.* Princeton Foundations of Contemporary Philosophy. Princeton, NJ: Princeton University Press, 2012.

Melia, Fulvio. *Cracking the Einstein Code: Relativity and the Birth of Black Hole Physics.* Chicago: University of Chicago Press, 2009.

——. *The Galactic Supermassive Black Hole.* Princeton, NJ: Princeton University Press, 2007.

Miller, Arthur I. *Empire of the Stars: Obsession, Friendship, and Betrayal in the Quest for Black Holes.* Boston: Houghton Mifflin, 2005.

Munns, David P. D. *A Single Sky: How an International Community Forged the Science of Radio Astronomy.* Cambridge, MA: MIT Press, 2013.

Musser, George. *Spooky Action at a Distance: The Phenomenon That Reimagines Space and Time—and What It Means for Black Holes, the Big Bang, and Theories of Everything.* 1st ed. New York: Scientific American/Farrar, Straus & Giroux, 2015.

Newton, Isaac. 1642–1727, *Principia.* English: *Sir Isaac Newton's Mathematical Principles of Natural Philosophy and His System of the World,* translated into English by Andrew Motte in 1729. Berkeley, California: University of California Press, 1934.

Overbye, Dennis. *Lonely Hearts of the Cosmos: The Story of the Scientific Quest for the Secret of the Universe.* Boston: Back Bay Books, 1999.

Penrose, Roger. *The Road to Reality: A Complete Guide to the Laws of the Universe.* London: Jonathan Cape, 2004.

Russell, Bertrand. *The ABC of Relativity.* London: K. Paul, Trench, Trubner, 1931.

Sagan, Carl. *Cosmos.* New York: Ballantine, 2013.

Susskind, Leonard. *The Black Hole War: My Battle with Stephen Hawking to Make the World Safe for Quantum Mechanics.* New York: Little, Brown and Co., 2008.

Taylor, Edwin F., and John Archibald Wheeler. *Exploring Black Holes: Introduction to General Relativity.* San Francisco: Addison Wesley Longman, 2000.

Templeton, John, and Robert L. Herrmann. *The God Who Would Be Known: Revelations of the Divine in Contemporary Science.* Philadelphia: Templeton Foundation Press, 1998.

Waller, William H. *The Milky Way: An Insider's Guide.* Princeton, NJ: Princeton University Press, 2013.

Weyl, Hermann. *Space, Time, and Matter.* Translated from the German by Henry L. Brose. London: Metheun & Co. Ltd., 1922.

Wheeler, John Archibald, and Kenneth William Ford. *Geons, Black Holes, and Quantum Foam: A Life in Physics.* 1st ed. New York: Norton, 1998.

Whitehead, Alfred North. *Science and the Modern World: Lowell Lectures, 1925.* New York: Macmillan, 1925.

Yau, Shing-Tung, and Steven J. Nadis. *The Shape of Inner Space: String Theory and the Geometry of the Universe's Hidden Dimensions.* New York: Basic Books, 2010.

JOURNAL ARTICLES

Almheiri, Ahmed, Donald Marolf, Joseph Polchinski, and James Sully. "Black Holes: Complementarity or Firewalls?" *Journal of High Energy Physics* 2 (2013).

Balbus, Steven A., and John F. Hawley. "A Powerful Local Shear Instability in Weakly Magnetized Disks. I-Linear Analysis. Ii-Nonlinear Evolution." *Astrophysical Journal* 376 (1991): 214–33.

Balick, B., and R. L. Brown. "Intense Sub-Arcsecond Structure in the Galactic Center." *Astrophysical Journal* 194 (1974): 265–70.

Bower, Geoffrey C., Heino Falcke, Robeson M. Herrnstein, Jun-Hui Zhao, W. M. Goss, and Donald C. Backer. "Detection of the Intrinsic Size of Sagittarius A* through Closure Amplitude Imaging." *Science* 304 (2004): 704–8.

Broderick, Avery E., and Abraham Loeb. "Imaging Bright-Spots in the Accretion Flow Near the Black Hole Horizon of Sgr A*." *Monthly Notices of the Royal Astronomical Society* 363 (2005): 353–62.

Doeleman, Sheperd, Eric Agol, Don Backer, Fred Baganoff, Geoffrey C. Bower, Avery Broderick, Andrew Fabian, et al. "Imaging an Event Horizon: submm-VLBI of a Super Massive Black Hole." In *astro2010: The Astronomy and Astrophysics Decadal Survey*, 2009.

Doeleman, Sheperd S., Jonathan Weintroub, Alan E. E. Rogers, Richard Plambeck, Robert Freund, Remo P. J. Tilanus, Per Friberg, et al. "Event-Horizon-Scale Structure in the Supermassive Black Hole Candidate at the Galactic Centre." *Nature* 455 (2008): 78–80.

Ekers, R. D., and D. Lynden-Bell. "High Resolution Observations of the Galactic Center at 5 Ghz." *Astrophysical Letters* 9 (1971): 189.

Falcke, Heino, Fulvio Melia, and Eric Agol. "Viewing the Shadow of the Black Hole at the Galactic Center." *The Astrophysical Journal* 528 (January 1, 2000): L13-L16.

Finkelstein, David. "Past-Future Asymmetry of the Gravitational Field of a Point Particle." *Physical Review* 110, no. 4 (1958): 965–67.

Genzel, R., and C. H. Townes. "Physical Conditions, Dynamics, and Mass

Distribution in the Center of the Galaxy." *Annual Review of Astronomy and Astrophysics* 25 (1987): 377–423.

Giddings, Steven B. "Black Holes and Massive Remnants." *Physical Review D* 46 (1992): 1347–52.

——. "Possible Observational Windows for Quantum Effects from Black Holes." *Physical Review D* 90 (2014).

Goss, W. M., Robert L. Brown, and K. Y. Lo. "The Discovery of Sgr A*." *Astronomische Nachrichten Supplement* 324 (2003): 497–504.

Hawking, S. W. "Black Hole Explosions?" *Nature* 248 (1974): 30.

——. "Information Preservation and Weather Forecasting for Black Holes." Submitted to arXiv.org January 22, 2014. https://arxiv.org/abs/1401.5761.

Imbriale, W. A. "Introduction to 'Electrical Disturbances Apparently of Extraterrestrial Origin'" 86, no. 7 (1998). doi:10.1109/JPROC.1998.681377.

Kellermann, K. I., and J. M. Moran. "The Development of High-Resolution Imaging in Radio Astronomy." *Annual Review of Astronomy and Astrophysics* 39 (2001): 457-509.

Lowe, David A., Joseph Polchinski, Leonard Susskind, Lárus Thorlacius, and John Uglum. "Black Hole Complementarity Versus Locality." *Physical Review D* 52, no. 12 (1995): 6997–7010.

Lynden-Bell, D. "Galactic Nuclei as Collapsed Old Quasars." *Nature* 223 (1969): 690–94.

——. "Searching for Insight." *Annual Review of Astronomy and Astrophysics* 48 (2010): 1–19.

Lynden-Bell, D., and M. J. Rees. "On Quasars, Dust and the Galactic Centre." *Monthly Notices of the Royal Astronomical Society* 152 (1971): 461.

Minkowski, Hermann. "Raum Und Zeit." *Physicalische Zeitschrift* 10 (1909): 11.

Oppenheimer, J. R., and H. Snyder. "On Continued Gravitational Contraction." *Physical Review* 56 (1939): 455–59.

Penrose, Roger. "Gravitational Collapse: The Role of General Relativity." *Nuovo Cimento Rivista Serie* 1 (1969).

Preskill, J., and D. V. Nanopoulos. "Do Black Holes Destroy Information?" In *Black Holes, Membranes, Wormholes, and Superstrings*, edited by Sunny Kalara, 1993: 22.

Ramesh, Narayan, Rohan Mahadevan, Jonathan E. Grindlay, Robert G. Popham, and Charles Gammie. "Advection-Dominated Accretion Model of Sagittarius A*: Evidence for a Black Hole at the Galactic Center." *Astrophysical Journal* 492, no. 2 (1998): 554.

Sanders, R. H. "The Case Against a Massive Black Hole at the Galactic Centre." *Nature* 359 (1992): 131.

Shakura, N. I., and R. A. Sunyaev. "Black Holes in Binary Systems. Observational Appearance." *Astronomy and Astrophysics* 24 (1973): 337–55.

Stanley, Matthew. "'An Expedition to Heal the Wounds of War': The 1919 Eclipse and Eddington as Quaker Adventurer." *Isis* 94, no. 1 (2003): 57–89.

Note on Endpapers

Astronomers actively debate the precise structure and position of the Milky Way's spiral arms, the density of stars and gas in the galactic bar, and the composition of the galactic center. The map used as endpapers reflects the most accurate information available at the time of publication. It is based on a 2008 illustration that the astronomer Robert Hurt created with infrared data from the Spitzer Space Telescope, and it incorporates adjustments based on newer research. The stylized speckle pattern accurately depicts our best understanding of the shape of the galaxy's spiral arms. The swarm of dots near the galactic center depicts the extraordinary density of stars found there. Within ten light-years of our sun, there are sixteen stars. Within ten light-years of Sagittarius A*, the black hole at the center of the galaxy, there are approximately a million stars. One percent of those stars—ten thousand dots—appear in the galactic-center inset on this map.

Index